Jaakko Astola
Pauli Kuosmanen

Fundamentals of
Nonlinear
Digital Filtering

CRC Press
Taylor & Francis Group
Boca Raton London New York

CRC Press is an imprint of the
Taylor & Francis Group, an **informa** business

CRC Press
Taylor & Francis Group
6000 Broken Sound Parkway NW, Suite 300
Boca Raton, FL 33487-2742

First issued in paperback 2019

ISBN-13: 978-0-367-44825-7 (pbk)
ISBN-13: 978-0-8493-2570-0 (hbk)

**Visit the Taylor & Francis Web site at
http://www.taylorandfrancis.com**

**and the CRC Press Web site at
http://www.crcpress.com**

Library of Congress Cataloging-in-Publication Data

Fundamentals of nonlinear digital filtering / edited by Jaakko Astola
and Pauli Kuosmanen.
    p. cm. — (Electronic engineering systems series)
  Includes bibliographical references and index.
  ISBN 0-8493-2570-6 (alk. paper)
  1. Signal processing—Mathematics. 2. Digital filters
(Mathematics) 3. Nonlinear theories. I. Astola, Jaakko.
II. Kuosmanen, Pauli. III. Series.
TK5102.9.F86 1997
621.382′2—dc21
                                                                97-10385
                                                                   CIP

Library of Congress Card Number 97-10385

ELECTRONIC ENGINEERING SYSTEMS SERIES

Series Editor: **J. K. FIDLER**, *University of York*

Asssociate Series Editor: **PHIL MARS**, *University of Durham*

To Our Children

Laura, Leena, Helena, and Pekka
Antti, Mikko, and Anna

Preface

The last few decades have changed the concept of nonlinear filtering into a full-grown field of signal processing with its characteristic features. The signal processing society now fully accepts nonlinear filtering as a necessity in certain application areas where linear filters cannot give satisfactory results. An introduction to the topic has clearly become necessary for every serious student of signal processing. Because understanding the delicate points in nonlinear signal processing requires some familiarity with the standard signal processing concepts, one semester course at the advanced undergraduate level would perhaps be ideal for introducing the topic. Books about linear systems and filters abound, while there are only a few books about nonlinear filtering, and they tend to be either encyclopedic-type ones or deep treatises on some particular subtopics. Our aim is to complement the present supply by fulfilling the need for a textbook about "the things that everyone should know". Our goal has been to write the textbook in a self-contained form so that it can be used for self-study and also as a handbook by engineers and scientists.

The one semester introductory course restriction has naturally tied our hands in a number of ways. The coverage of a book of reasonable length can not be exhaustive while it should be rigorous; the book should not be stodgy while it should provide enough information in a digestible form; the book should be about widely acknowledged fundamentals and not about personal preferences; the book should be about seeing interrelationships, connections, and integrating one's knowledge, not about fragments of ideas with no unification. How do we do this, since the task is extremely challenging? By consistently reviewing pertinent concepts and relating them together, and by unifying notations, deleting overlaps and discovering a common foundation where most of the results are deduced from. By keeping the exposition as simple as possible and by injecting mathematical methods in a balanced way; large doses where that is the only way to get strong results and the patient would not function well otherwise and small doses where the side effects of extra formalism are more harmful than the seeming rigor.

We start by discussing the problems that should be solved by filtering and carefully analyze the drawbacks of the basic solutions, the mean and the median filters. We also carefully explore the intimate connection between filtering and statistical estimation. Because of this important connection knowledge of statistics enables one to understand the sometimes very complicated behavior of nonlinear filters. Chapter 2 builds up the necessary operational issues of statistics. Chapter 3 provides background information and motivation for different nonlinear filtering approaches designed to overcome the drawbacks of the basic solutions. For every filter we aim to provide the information that is needed to understand for what purpose the filter is meant, how the purpose can be achieved by this filter, and how it actually operates. Around 30 different filter classes have been studied by going through their fundamentals: ideas, definitions, algorithms, and impulse and step responses. The notations used for different filter classes are unified, since the variety of notations in publications easily causes a degree of consternation. We extend the understanding of the filtering operations by checking their results by unusually detailed examples. These examples make it possible, for the first time in the course of the history of nonlinear filtering, to directly compare the differences of filtering operations.

We have intentionally kept Chapter 3 long and not split it into several chapters each containing filters that are related to each other in some sense. This is because there are several equally justifiable groupings of filters depending on the viewpoint. This becomes clear in Chapter 3 where we provide examples of such groupings.

The nature of Chapter 4 is very theoretical, but still it provides tools than can be directly applied in real filter design problems. We have chosen three major classes of nonlinear filters to illustrate what an exact statistical analysis and optimization typically require.

Problems are supplied at the end of the book. Many of the problems are meant to test and consolidate the presented work, while some of them also contain theoretical material which is useful in the analysis of the filter behavior.

We hope that after reading this book the reader feels that we have succeeded in doing our task and in putting the ideas behind nonlinear filtering together in ways that support learning. Improvements and corrections to our presentation are likely and welcome.

It is hoped that this book will be an appetizer, whetting one's appetite for more reading. This book is about fundamentals, and there is still much more to come. Nonlinear filtering is habit forming. Our best reward for this book is to help nonlinear filters going by addicting new talented researchers and application engineers to the discipline.

How to avoid reading the book

This is easy, just stay stubborn and keep on lying to yourself that there is no need for nonlinear filters. This is partially true, since there exist many filtering applications where there is no need to consider nonlinear methods. Still, let us see the other side of the linearity. Actually, linearity is a constraint! We understand nonlinearity as "not necessarily linear", and thus allowing the use of nonlinear techniques, the linearity constraint is removed, and obviously, the obtained results cannot be any worse than with the requirement. Secondly, we would like you to recall that the present knowledge of the human visual system indicates that it possesses nonlinear characteristics. This should be taken into account in image filtering. All image processing is—of necessity—nonlinear.

How to read the book

A quick overview of the topic is achieved by going through Chapter 1 and then the ideas behind different filters presented in Chapter 3. To help the reader to localize these ideas from the middle of the body text they are marked in the margin.

For a practically oriented reader the best way to avoid reading the whole book is to go through Chapters 1 and 3. Some Sections in Chapter 3 are easier to understand if one consults Sections 2.2 and 2.3 while reading Chapter 3. In particular, Section 3.11 and Section 3.12 rely heavily on ideas from estimation theory.

Those who know statistics rather well can just browse through Sections 2.1 and 2.3 of Chapter 2 in order to familiarize themselves with the notation used in the book. The material in Section 2.4 is slightly nonstandard and is needed mainly in Chapter 4. Even though Chapter 2 is, in principle, self-contained, it is quite terse. Thus, for those who are not familiar with most of the concepts, it will be helpful to consult an introductory book on probability theory and its applications in order to obtain an intuitive understanding of the ideas behind the concepts.

The mathematical complexity is highest in Sections 2.3, 2.4 and in Chapter 4. In every chapter the key ingredient in the learning process is an active involvement of the reader. You are strongly encouraged to solve the exercises, as they contain important pieces of information about the filters. Furthermore, we encourage you to be skeptical about the presented approaches. None of them is perfect, and to realize this indicates that our message has gone through.

For those who are interested in solid theoretical considerations Chapter 4 provides a good starting point. It contains a brief introduction to the analysis and optimization methods that have been developed for the filter classes that are based on order statistics. Though necessarily very incomplete, the chapter is intended to give a clear picture of the "state of the art" in this particular research field.

Historical notes

Nonlinear filtering has been considered even in the fifties [113]. Since then, the field has seen a rapid increase of interest indicated, e.g., by the ever increasing number of publications. A brief look at the bibliography reveals that most of the filters studied in this book have been introduced in the eighties. We have not tried to trace the first occurrence of the ideas in the literature and by no means claim that the references we provide give honor to the right persons. Our references follow mainly the "western"-line. Still it is evident that many of the studied filters have been introduced earlier, e.g., in eastern countries. For example, the L-filters, multistage medians and median hybrid filters have been rather extensively studied from the theoretical point of view in the beginning of the seventies in the Soviet Union (see [28] and references therein) and have been independently reinvented and put into wide practical use around 15 years later by western researchers. To find out the exact inventors for every filter would require an enormous amount of work and it is left to another study, as it is definitely not needed in a textbook.

Acknowledgments

Writing a book is a humbling experience—to realize how dependent one actually is upon other people.

We would like to express our gratitude to our family members for bearing with us during the preparation of this book and providing the support and encouragement needed daily.

Special thanks are due to Mr. Kai Willner, Mr. Heikki Huttunen, Mr. Petri Granroth, and Ms. Sari Siren, who have implemented the algorithms herein. Especially, we are indebted to Mr. Kai Willner who spent many overtime hours generating the image illustrations and forbearing with our constant changes of mind.

We wish to thank our friends, colleagues, and students—Dr. Karen Egiazarian, Ms. Katriina Halonen, Mr. Heikki Huttunen, Mr. Pertti Koivisto, Mr. Ronan Mac Laverty, Prof. Sanjit K. Mitra, Prof. Yrjö Neuvo, Prof. Ioannis Pitas, and Prof. Tapio Saramäki for providing us pages of comments, constructive suggestions, and corrections.

Contents

List of Symbols and Abbreviations

Signals and Systems

$\mathbf{s} = (S_1, S_2, \ldots, S_n)$	digital noise-free signal
$\mathbf{x} = (X_1, X_2, \ldots, X_n)$	digital noisy signal
$\mathbf{n} = (N_1, N_2, \ldots, N_n)$	digital noise signal
X_i	input sample
$\mathbf{x} = (X_1, X_2, \ldots, X_N)$	data vector
X^*	the center sample, which is being filtered
$X_{(i)}$	ith order statistic, i.e., the ith smallest value
$R(X_i)$	rank of the sample X_i
$\mathbf{a} = (a_1, a_2, \ldots, a_N)$	weight vector

Probability

Ω	sample space
$\mathcal{F}$	σ-algebra
P	normed measure
$(\Omega, \mathcal{F}, P)$	probability space
$P(A)$	probability of the event A
X, Y	random variables
x, y	realizations of random variables X and Y
$F_X(\xi)$	distribution function of X
$f_X(\xi)$	density function of X
$\mu = E\{X\}$	expectation of X
$\mu'_k = E\{X^k\}$	kth moment of X
$\mu_k = E\{(X - \mu'_1)^k\}$	kth central moment of X
$\mu_2 = \sigma^2 = \text{Var}\{X\}$	variance
σ	standard deviation
$N(\mu, \sigma^2)$	normal/Gaussian distribution (mean μ, variance σ^2)
$U(a, b)$	uniform distribution (mean $\frac{(a+b)}{2}$, variance $\frac{(b-a)^2}{12}$)
$F(\xi_1, \xi_2, \ldots, \xi_N)$	joint distribution function of $X_1, X_2, \ldots, X_N$
$f(\xi_1, \xi_2, \ldots, \xi_N)$	joint density function of $X_1, X_2, \ldots, X_N$
$\boldsymbol{\mu} = (\mu_1, \mu_2, \ldots, \mu_N)^T$	mean vector of $X_1, X_2, \ldots, X_N$
$P(A|B)$	conditional probability of A assuming B
$f(\xi|y)$	conditional density of X assuming $Y = y$
$E\{X|y\}$	conditional expectation of X assuming $Y = y$
$T(y)$	regression line
$f(x_0)dx$	probability element
$f_{\mathbf{x}}(x_1, \ldots, x_N)dx_1 \cdots dx_N$	probability element
$T(X_1, X_2, \ldots, X_N)$	estimator (statistic)
$\theta_1, \theta_2, \ldots, \theta_k$	parameters defining a distribution function
$\boldsymbol{\theta} = (\theta_1, \theta_2, \ldots, \theta_k)$	parameter vector
$F(\cdot; \boldsymbol{\theta})$	distribution function with parameter vector $\boldsymbol{\theta}$
$f(\cdot; \boldsymbol{\theta})$	density function with parameter vector $\boldsymbol{\theta}$
Θ	parameter set

$L(\boldsymbol{\theta}: X_1, X_2, \ldots, X_N)$ likelihood function
$\hat{\boldsymbol{\theta}}(\mathbf{x}), \hat{\boldsymbol{\theta}}$ estimate of the parameter vector θ
$\arg\min_{\theta \in \Theta} f(\theta)$ the value of the argument $\theta \in \Theta$ minimizing $f(\theta)$
$D(\alpha_1, \alpha_2, \ldots, \alpha_k; \beta)$ k-dimensional Dirichlet distribution
$D^*(\gamma_1, \gamma_2, \ldots, \gamma_k; \delta)$ ordered k-dimensional Dirichlet distribution
MAE   mean absolute error
MSE   mean square error
i.i.d.   independent and identically distributed
MLE   maximum likelihood estimator
$\Phi_i(t)$ distribution of input sample X_i
$\Phi(t)$ distribution of i.i.d. input
$\Psi(t)$ output distribution

Functions

$\delta(x)$ delta function
$\text{sign}(x)$ sign function
$\rho(x), \psi(x)$ functions defining M-estimators/filters
$\Gamma(x)$ Gamma function
$N!$ N factorial $(N! = 1 \cdot 2 \cdots N)$
$f(x_1, x_2, \ldots, x_N)$ Boolean function of the variables $x_1, x_2, \ldots, x_N$
$T^{(j)}(x(i))$ threshold decomposition function

Miscellaneous symbols

$\mathbf{R}$ set of real numbers
$\mathbf{Z}$ set of integers
$\mathbf{Z}_+$ set of positive integers
$\mathbf{x}^T$ transpose of the vector $\mathbf{x}$
$\mathbf{C}, \mathbf{R}, \mathbf{Q}, \mathbf{J}$ matrices
$|\mathbf{C}|$ determinant of the matrix $\mathbf{C}$
$\mathbf{C}^{-1}$ inverse of the matrix $\mathbf{C}$
$\lceil \sigma_1^2, \sigma_2^2, \ldots, \sigma_N^2 \rfloor$ diagonal matrix with $\sigma_1^2, \sigma_2^2, \ldots, \sigma_N^2$ on the diagonal
$\mathbf{0} = (0, 0, \ldots, 0)^T$ zero vector
$\mathbf{1} = (1, 1, \ldots, 1)^T$ unity vector
sup   supremum
MAD   median of absolute deviations from the median
MAX   maximum
MIN   minimum
MED   median $(X_{(k+1)}$, when $N = 2k + 1)$
$\emptyset$ empty set
$r \Diamond x$ repetition (duplication) operation, x repeated r times
$O(g(n))$ asymptotically dominated by $g(\cdot)$, $f(n) = O(g(n))$ if and only if there exist real numbers A and n_0 such that $|f(n)| \leq A|g(n)|$ for all $n > n_0$
$w_H(\mathbf{x})$ Hamming weight of the binary vector $\mathbf{x}$

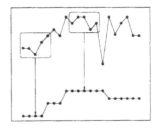

1

NONLINEAR SIGNAL PROCESSING

NONLINEARITY BEGETS COMPLETENESS;
MISJUDGEMENT CREATES LINEARITY
Ch. XXII Lao Tzu (Circa 600 BC)

1.1 Signal Processing Model

A (physical) *signal* can be defined as the physical carrier of information. There are numerous familiar examples of signals. Audio signals are variations in air pressure and sensed by the ear. Visual signals are light waves carrying information of a scene to the light sensing cells of the retina. A typical signal in communication theory is the varying voltage measured across two nodes of an electric system. Another typical example of a signal would be a black-and-white photograph, where the information is represented as the varying lightness of the photographic paper.

A signal can be mathematically represented in many ways. A natural representation of a time-varying voltage is a function $f(t)$, where the independent variable t denotes time. The same formulation suits audio signals equally well. The representation of light waves carrying information from a scene to the eye is quite complicated but the final image is simply represented as a function of two variables $f(x, y)$. For a black-and-white image, (x, y) is the spatial location of the point in the image plane and $f(x, y)$ is the brightness value of that point. A color image is represented as a vector-valued function

1

$(R(x, y), G(x, y), B(x, y))$, where R, G, and B are the intensities of the red, green, and blue colors (the so called primaries), respectively. Other representations such as luminance, hue, and saturation are also possible.

We can make a crude division of signal processing tasks into the following three categories:

- *Removal of interference.* Typical examples would be the filtering of noise or channel distortion in communication systems and the removal of noise from images. This type of processing is often viewed as the estimation of the underlying signal.

- *Transformation of signals into some other form more suitable for particular purposes.* Typical examples would be digitizing a photograph or an analog audio signal, modulation of a communication signal to a different frequency band, taking a Fourier transform of a digital signal to analyze its frequency content, and data compression.

- *Analysis and extraction of some characteristics of the signal.* Typical examples could be an EEG or EKG signal analysis in medical sciences, or speech recognition in communications as well as object recognition in computer vision. Taking a Fourier transform of a signal could also belong to this category.

This book concentrates mostly on the first task. It is also very often a necessary preprocessing stage before the actual processing stage. The problem we are mainly addressing is the following. We observe a signal (one- or two-dimensional) that is corrupted by some kind of noise. The noise may be additive, multiplicative or even of a more general nature, and some assumptions are made about its statistical characteristics. Based on the observed signal, our task is to estimate the original uncorrupted signal as accurately as possible.

We define the signals as functions of continuous variables, usually t representing time for one-dimensional signals, and (x, y) representing spatial coordinates in the image plane for two-dimensional signals. We consider the digital processing of signals, and so their representation as functions of continuous variables must be transformed to representations as functions of discrete variables. There are profound results about the conditions under which this transformation can be done with no loss of information. We will not go into details and only assume that this digitization has been successfully performed.

We present one-dimensional signals as sequences of real numbers or, equivalently, as real valued functions of integers $f : \mathbf{Z} \to \mathbf{R}$. Likewise, gray-scale two-dimensional signals or images are represented as two-dimensional sequences of real numbers, i.e., functions $f : \mathbf{Z} \times \mathbf{Z} \to \mathbf{R}$. Thus, it should now be clear that we will not consider color images at all in this book. In practice the signals cannot be of infinite extent and we assume $f(n)$ to be defined only for $1 \leq n \leq A$ for one-dimensional signals and $f(m, n)$ to be defined only for $1 \leq m \leq A, 1 \leq n \leq B$ for images. Another alternative is to set f equal to zero outside the above ranges.

As the signals are to be processed by digital devices, the values must be quantized to a finite number of values. The number of quantization levels depends on the application. For one-dimensional signals, the number of quantization levels can be as high as 24 bits, and the computations can be done in floating point arithmetic. For images viewed by humans there is seldom a need for more than 256 intensity levels.

1.2 Signal and Noise Models

In this book we will concentrate mostly on nonlinear filtering techniques for removing an unknown corruption from a signal. The most general setting for the problem is the following. There is a signal $\mathbf{s} = (S_1, S_2, \ldots, S_n)$ that is generated by some system. It is a random process, and we have some information about the nature of the process. For instance, we may know the probability density function of the signal. The signal is corrupted by a likewise random mechanism into another signal $\mathbf{x} = (X_1, X_2, \ldots, X_n)$. Again, we make some assumptions about the corruption that is modeled by the conditional probability density $P(\mathbf{x}|\mathbf{s})$. This can be interpreted as the probability that we observe the signal $\mathbf{x}$ if the actual signal was $\mathbf{s}$. Our task is to find the original signal $\mathbf{s}$ based on the observed signal $\mathbf{x}$.

Of course, in such a general setting there is no hope of finding a useful solution. However, by making simplifying assumptions we may be able to reduce the problem to a form that can be solved using methods that are available. The usefulness of the solution for the reduced problem, of course, depends on the validity of the assumptions. The complexity of the simplified problem depends mainly on the following three factors:

- The mechanism generating the underlying signal $\mathbf{s}$, i.e., the model for the underlying signal;

- The nature of the corruption;

- The measure of the accuracy of the solution with respect to the above assumptions.

The beautiful theory of linear filtering gives optimal methods when the corruption can be modeled as a Gaussian process and the accuracy criterion is the mean square error. These assumptions are reasonable in most applications, but there are also cases when they cause serious problems. In digital systems, errors are often caused by bit changes, and the resulting distribution is far from Gaussian. In image processing, where the visual quality is often the ultimate criterion, the mean square error is not a realistic criterion. An image with a large mean square error can be visually much better than another with a small mean square error. It is in those cases where the assumptions of the classical linear theory are violated that the nonlinear methods prove most useful.

1.3 Fundamental Problems in Noise Removal

In the rest of this chapter we illustrate the fundamental problems related to noise models and filtering methods by means of illustrative examples. They are simple examples but still reveal the main difficulties that we face in the search for better methods in signal and image processing.

We shall compare two filters, the *mean filter* and the *median filter*, for a few simple cases. The purpose of the filtering operation is assumed to be an effective elimination or attenuation of the noise that is corrupting the example signals. We consider both one-dimensional and two-dimensional (image) cases. It is easier to understand what is going on with the one-dimensional signal. On the other hand, the effects are better visualized with images.

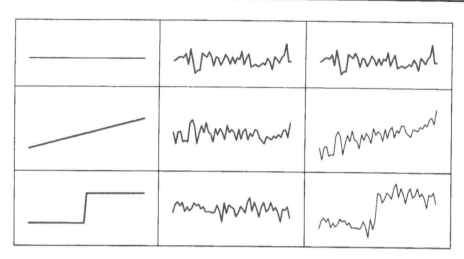

Figure 1.1. Three experimental one-dimensional signals (left column), noise (middle column), and the corrupted signal, i.e., signal+noise (right column).

1.3.1   One-Dimensional Signals Corrupted by Additive White Gaussian Noise

Consider the following three signal segments: a constant signal, a ramp signal or linear slope, and a step signal. Each of these is corrupted by additive Gaussian noise, and we compare the effects of the mean and the median filters of the same window size.

Consider the following signal:

$$x(i) = s(i) + n(i), \quad i = 1, 2, \ldots, 50,$$

where

- $s(i) = a, \quad i = 1, 2, \ldots, 50$, (constant) in the first case;

- $s(i) = 0.2i, \quad i = 1, 2, \ldots, 50$, (ramp) in the second case;

- $s(i) = \begin{cases} 0, & i = 1, 2, \ldots, 25, \\ 10, & i = 26, 27, \ldots, 50, \end{cases}$ (step signal) in the third case.

We assume that $n(i)$, $i = 1, 2, \ldots, 50$, are independent Gaussian random variables with mean equal to 0 and variance equal to 4. These signals are illustrated in Figure 1.1, where the signals are shown in the left column, noise in the second column, and the corrupted signals in the third column.

Let us filter these signals with a mean filter of length seven. In general, the one-dimensional mean filter of length $N = 2k + 1$ is defined by specifying its output at time instant i as

$$y(i) = \frac{1}{2k + 1} \sum_{j=-k}^{k} x(i + j).$$

The range of indices $i - k, i - k + 1, \ldots, i + k$ over which the mean is computed is usually called the *filter window*. Sometimes these types of filtering operations are called *moving window filters*. The signal samples in the window at time instant i are $x(i-k)$, $x(i-k+1)$,

..., $x(i + k)$, and the filtering operation can be more general than the mean operation. For the filter of length seven the output signal $y(i)$ equals

$$y(i) = \frac{1}{7} \sum_{j=-3}^{3} x(i+j), \quad i = 4, 5, \ldots, 47.$$

The first, obvious but very important, observation is that the mean filter is a *linear operation*. Recall that a mapping $f : A \to B$, where A and B are (complex) linear spaces, is called linear if

$$f(\alpha a + \beta b) = \alpha f(a) + \beta f(b)$$

for all $a, b \in A$ and complex numbers α and β. This observation means that the action of the filter on the uncorrupted signal $s(i)$ and the noise signal $n(i)$ can be investigated separately:

$$y(i) = \frac{1}{7} \sum_{j=-3}^{3} x(i+j)$$

$$= \frac{1}{7} \sum_{j=-3}^{3} s(i+j) + \frac{1}{7} \sum_{j=-3}^{3} n(i+j)$$

$$= (\text{filtered underlying signal}) + (\text{filtered noise}).$$

The results of this filtering operation are illustrated in Figure 1.2.

From statistics we know that the variance of the mean of seven independent random quantities, each having variance $= 1$, equals $\frac{1}{7}$. Thus the overall result is that the noise power is reduced by 86% (or 8.5 dB in engineering parlance), but as the result there will also be distortion which is caused by filtering the underlying signal by the mean filter.

Exact analysis of the effect of the mean filter on different types of signals is easily carried out by using the frequency domain representation (Exercise). For our purposes, it suffices to say that, in practice, a signal that looks like a straight line can pass the mean filter unaltered and that the changes caused by the filter are most pronounced if the signal has sharp corners or discontinuities.

Consider, then, filtering the same signals by a median filter of length seven. This filtering operation is given by

$$y(i) = \text{MED}\{x(i-3), x(i-2), x(i-1), x(i), x(i+1), x(i+2), x(i+3)\}, \quad i = 4, 5, \ldots, 47,$$

where MED takes the center value after ordering (e.g. $\text{MED}\{2, 7, 1, 6, 5, 3, 4\} = \text{MED}\{1, 2, 3, 4, 5, 6, 7\} = 4$), and the results of the filtering operation are illustrated in Figure 1.3. The median filter is attributed to Tukey [105, 104].

The first observation is that, unlike for the mean filter, it is not possible to apply the operation to the signal and the noise separately. Equivalently, we say that the filter is not linear. The second observation is that because of this impossibility of separating the action on the signal and the noise, the concept of noise power attenuation is not well defined, even for independent Gaussian noise. We can only investigate the reduction in noise power for each underlying signal separately.

The third observation is that all the noise-free (underlying) example signals $s(i)$ are completely preserved by the median filter. This fact is extremely important in image processing because from the properties of the human eye, it follows that edges often carry the main information in an image.

When examining the effect of the median filter on the corrupted (noisy) signals there are three noticeable facts. First, the noise on the constant signal appears to be attenuated

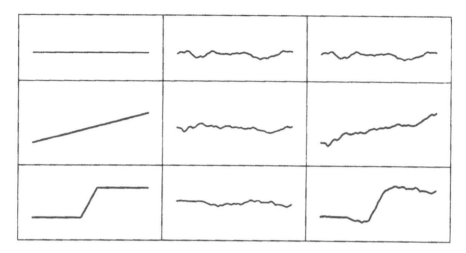

Figure 1.2. Mean filtering (window length 7) of the experimental one-dimensional signals. The underlying signal $s(i)$ filtered (left column), the noise signal $n(i)$ filtered (middle column), and the corrupted signal, i.e., signal+noise $s(i) + n(i)$ filtered (right column).

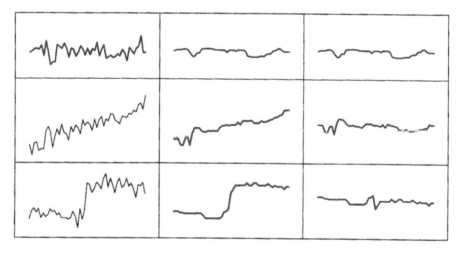

Figure 1.3. Median filtering (window length 7) of the experimental one-dimensional signals. The corrupted signal $s(i) + n(i)$ (left column), the filtered corrupted signal (middle column), and the difference between the filtered signal and the original signal (right column).

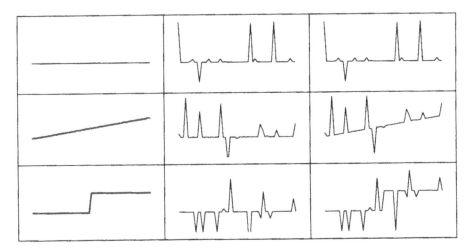

Figure 1.4. Three experimental one-dimensional signals (left column), noise (middle column), and the corrupted signal, i.e., signal+noise (right column).

to the same degree as with the mean filter. This fact can also be easily demonstrated analytically.

Second, the median filter does not attenuate noise well in the case of the noisy ramp signal. It tends to form a staircase-like pattern. This is intuitively quite clear because if the ramp is steep enough, additive noise will not change the order of the samples in the window of the median filter.

Third, the noisy step signal is markedly rounded at the step. This is also intuitively clear. If the center of the median filter is at the lower end of the step edge, then because there are three samples of the window on the upper part of the step, the median is essentially the maximum of the four samples on the lower part within the window.

These properties can be seen in Figure 1.3, where in the left column there are the noisy constant, ramp and step signals, and in the middle column are the corresponding filtered signals. The right column shows the differences between the filtered signal and the original. From these difference figures we see how strongly the underlying signal affects the noise attenuation of the median filter.

1.3.2 One-Dimensional Signals Corrupted by Impulsive Noise

We repeat our experiment in the preceding section for impulsive noise. The signals and the noise are shown in Figure 1.4 and the filtering results are shown in Figures 1.5 and 1.6 for the mean and the median, respectively.

We can immediately observe several differences in the behavior of the mean filter and the median filter for the impulsive noise. In the case of a single impulse, the mean filter spreads the impulse but also reduces the amplitude, whereas the median filter totally eliminates the impulse. In fact the median filter with window length $N = 2k+1$ completely removes impulses of length below $k+1$ independent of their sign. This is further illustrated in Figure 1.7.

This is a fundamental difference between the behavior of a linear and a nonlinear filter. The *impulse response* (that is, the output signal when the input signal is equal to 1 at

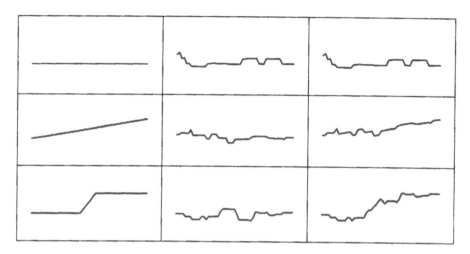

Figure 1.5. Mean filtering (window length 7) of the experimental one-dimensional signals. The underlying signal $s(i)$ filtered (left column), the noise signal $n(i)$ filtered (middle column), and the corrupted signal, i.e., signal+noise $s(i) + n(i)$ filtered (right column).

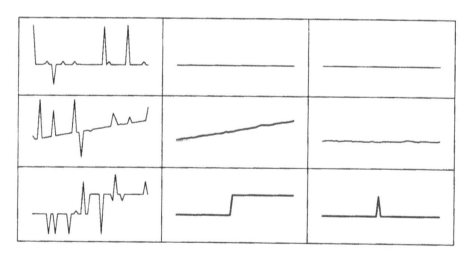

Figure 1.6. Median filtering (window length 7) of the experimental one-dimensional signals. The corrupted signal $s(i) + n(i)$ (left column), the filtered corrupted signal (middle column), and the difference between the filtered signal and the original signal (right column).

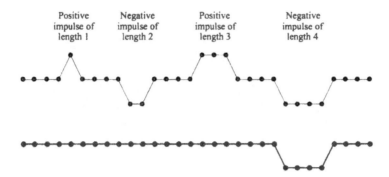

Figure 1.7. The median filter and impulses (window length 7). All impulses of length less than 4 are completely removed by the filter.

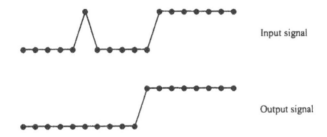

Figure 1.8. Edge jitter in median filtering (window length 11); edge is shifted because of an impulse near it.

time instant $i = 0$ and zero otherwise) of a nontrivial time-invariant linear filter clearly cannot be a zero-valued sequence (Exercise).

It is important to notice that this ideal behavior of the median filter holds only for the case where we have an impulse on a constant signal. For example, if the impulse is close to an edge it may be removed but at the same time the edge moves towards the impulse. This is called *edge jitter* and is illustrated in Figures 1.6 and 1.8.

To further analyze the noise attenuation of the mean and the median filters in one-dimensional signals, we apply them to an artificial one-dimensional signal containing typical important signal fragments like steps, ramps, oscillations, and impulses of different widths. First, we evaluate the filtering results when the filter is directly applied to the noise-free signal. This gives us information about how well the filter can preserve details or, more precisely, which details will be removed or altered. Second, we evaluate the filtering results when the filter is applied to the noisy signal, where the noise is additive Gaussian noise together with impulsive noise caused by random bit errors. The results are shown in Figures 1.9 and 1.10. The top row ((a) and (b)) shows the original signals (dotted line) and the filtering results (solid line), the second row ((c) and (d)) shows the error signal, i.e., the difference between the filtering result and the noise-free signal.

We encourage the reader to carefully analyze these figures as we consider their information content very large. Consider first the noise-free case. Both of the filters create the largest error in the beginning of the signal where the impulses were located. In the rest of the signal, the median filter is superior to the mean filter as it removes only the sharpest details of the signal and leaves the rest untouched. The mean filter also causes

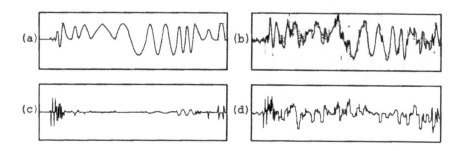

Figure 1.9. One-dimensional signals filtered by the mean filter with window length 11.

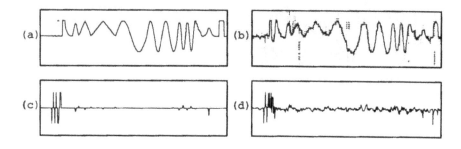

Figure 1.10. One-dimensional signals filtered by the median filter with window length 11.

other detail losses to the signal. Most notably, the pulse in the end of the signal is changed quite dramatically. The reason is obvious: the pulse is formed actually by two consecutive step edges, which are blurred by the mean filter. Consider then the noisy signal filtering. Again there is no doubt which filter performs better. The original signal shape is quite well preserved in the median filtered signal but not in the mean filtered signal. However, the median filter cannot remove the Gaussian noise from the signal. This causes the error seen in Figure 1.10 (d). The error seen in Figure 1.9 (d) is mainly due to the impulses. This is easily seen by comparing the subplots (b) and (d).

1.3.3   Mean and Median in Image Filtering

Consider next the differences that occur when filtering images using the mean and the median filters with the 5×5 square window. In a manner similar to the one-dimensional case, the moving window is now placed at every pixel, the operation (mean/median) uses the 25 pixel values inside the window, and the result of the operation is the output at the window location.

Two example images are used. The first one, "Geometrical" (see Figure 1.11), is artificially created, and its structure is quite regular. In this image there are certain structures that one often, but not always, wants to preserve. Step edges in the chessboard background and thin lines are such ones. The zone-plate in the right hand side of the "Geometrical" image is very frequency oriented. It can be used to study the manner in which different frequencies are amplified/attenuated by filtering. The results after the mean and median filtering are shown in Figures 1.12 and 1.13, respectively.

Figure 1.11. Image "Geometrical".

Figure 1.12. "Geometrical" filtered by the mean filter with window size 5 × 5.

Figure 1.13. "Geometrical" filtered by the median filter with window size 5 × 5.

The differences between these two filters are again clear. As the mean filter is not able to preserve steps, and the image "Geometrical" contains dozens of them, the overall visual appearance after mean filtering is very fuzzy or blurry. The result is far away from a pleasure to the eye. But the same holds for the median filter as well. Even though the result is sharp, many important details are lost. Some details like thin lines are not only lost but their original place can be somehow disturbingly discerned in the result. These lines can be seen from the mean filtered image even though they are blurry. Still, visually the median filtered image is preferable to the mean filtered one.

From the frequency part of the image "Geometrical" we can also see that both filters act like low-pass filters as the high frequencies are attenuated more than the low frequencies.

The second original image under investigation is shown in Figure 1.14. The image contains many details that cause difficulties in filtering. It has smooth regions with fine texture, small details, sharp edges, and regular patterns. On the right hand side there are four enlarged subimages with characteristic features. Different parts cause problems for different image processing algorithms. The irregular details in the hat and the scarf (first and third subimages) are difficult for any algorithm that has a strong noise attenuation capability. We shall see that the regular part on the background (second subimage) will be problematic to many filters. The human eye is very good in recognizing even small distortions, especially in a human face. The fourth subimage was chosen to illustrate the distortions the filters cause in images of human faces.

Right below the image is a scan line from the lower part of the image. The location of the line is shown in the large image by small marks on both sides of the image. The scan line shows, e.g., the blurring caused by some filters quite well.

In this example we add impulsive noise to the image. The impulsive noise is generated by introducing random bit errors (with probability 0.05) to obtain a more realistic case than the often used "pure" salt-and-pepper noise with amplitudes 0 and 255.

In the noisy image (Figure 1.15) and in the processed images (Figures 1.16 and 1.17) on the extreme right there are shown the difference images taken between the original and processed subimages. The intensities of the difference images are multiplied by two to make them more visible. Sometimes algorithms cause a systematic change or shift in a detailed part. This may not be visible from the image itself but is clearly shown in the difference image. Systematic shifts are not so bad if images are processed for visual viewing, but if we are preprocessing images for machine vision applications systematic shifts can cause difficulties.

Below the scan line is the histogram of the absolute difference image between the original and the processed image. Careful analysis of this histogram can reveal quite a lot about the behavior of the filter. Moreover, the overall (empirical) mean absolute error (MAE) and the overall mean square error (MSE) are given.

We examine several particular problems in image filtering using this example image. Remember that our main objective is to filter out the noise in such a way that the original image is restored while paying special attention to the visual quality of the restored image. There is no reliable quantitative measure of success for this task. For example, an image with very annoying artifacts may have small mean square or mean absolute errors. In this simple and restricted case, we can visually evaluate the performances of the mean and the median filters regarding the noise attenuation, the overall visual quality, the detail preservation, and the edge preservation.

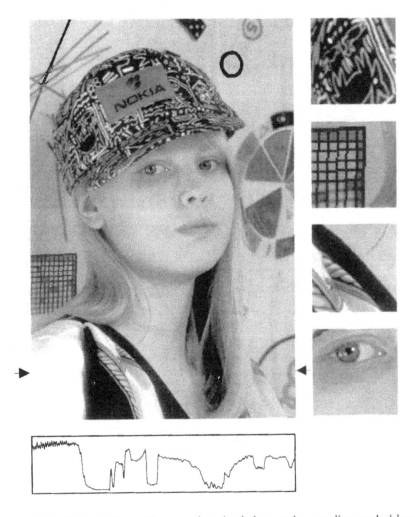

Figure 1.14. Original "Leena" image, selected subplots, and a scan line marked by arrows.

Noise Attenuation

When comparing the histograms of the absolute difference images (between the filtered and the noise-free image) by the mean filter (Figure 1.16) and the median filter (Figure 1.17), as well as the numerical measures of noise attenuation, it is seen that both filters attenuated the noise, especially in the mean square error sense. The median filter attenuated the noise better than the mean filter. Especially, the number of error-free values is much larger in the median filtered image. This is natural because the mean filter tends to change almost all pixel values, whereas for the median filter the output is always one of the values in the window and so an error-free value has a greater chance to survive the filtering.

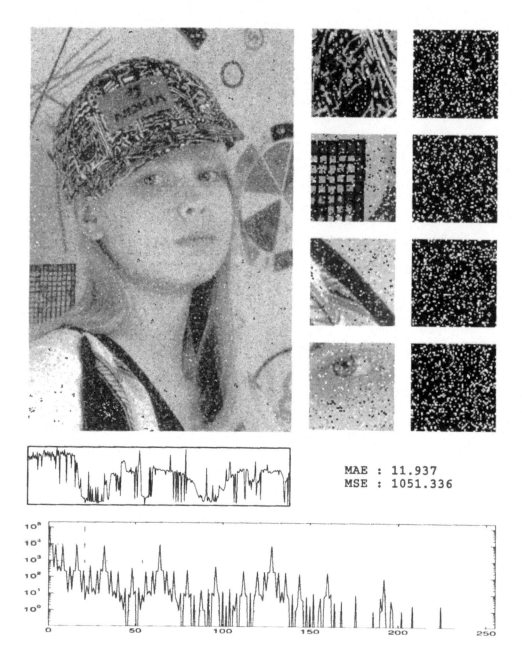

MAE : 11.937
MSE : 1051.336

Figure 1.15. "Leena" with noise, selected subplots and their difference image from the noise-free image, a scan line, the histogram of the absolute error, and the numerical measures: the mean absolute error (MAE) and the mean square error (MSE).

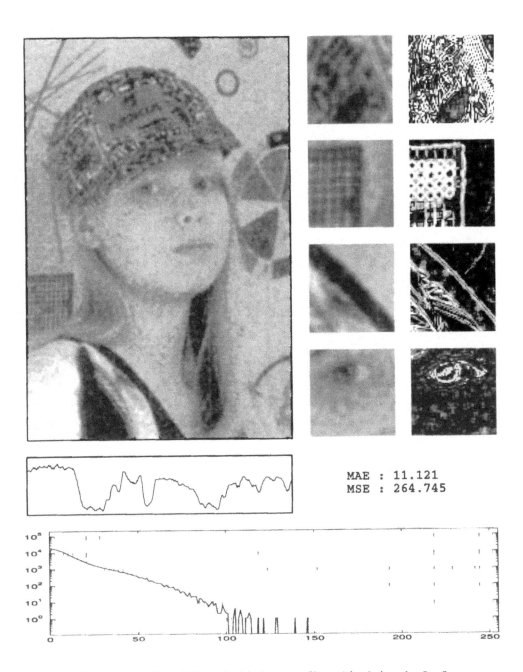

MAE : 11.121
MSE : 264.745

Figure 1.16. "Leena" filtered with the mean filter with window size 5 × 5.

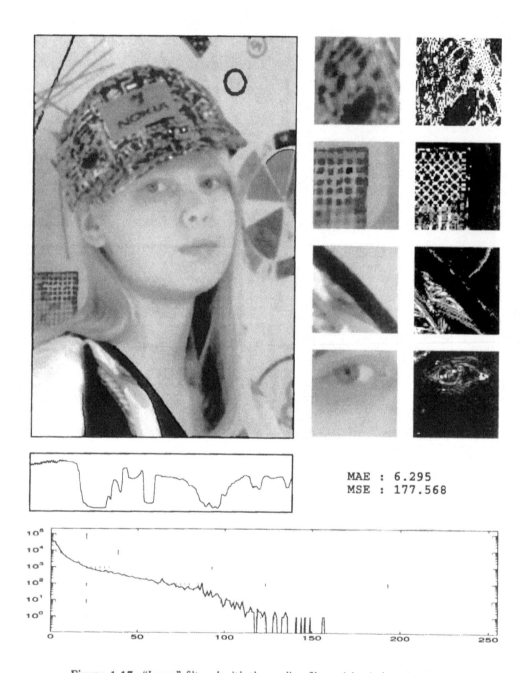

MAE : 6.295
MSE : 177.568

Figure 1.17. "Leena" filtered with the median filter with window size 5 × 5.

Overall Visual Quality

Once one looks at the images themselves, it is hard to believe that the numerical measures of noise attenuation are relatively similar because the median filtered image looks markedly better. The reason is that the remaining error is spread differently over the image for the mean and the median filters. The mean filter leaves a small annoying blob at each impulse, while the median filter appears to handle the impulses very well.

Detail Preservation

In preserving details both of the filters have quite serious problems. The comparative performances are best shown in the enlarged detail and difference images. Consider the first enlargement images. Obviously, the original pattern of the hat is best visible in the noisy image and more or less wiped out by both filters. The fact that the original image has a high frequency content causes even the impulsive noise to be visually acceptable in this image. Looking at the logo in the hat, it is seen that the median filter handles larger details slightly better. The second enlargement image reveals a very interesting property of the median filter. The regular pattern is effectively transformed into the negative of the original even though the noise is well removed. This phenomenon is related to the well known fact [5, 107] that the median filter passes certain high frequency signals unattenuated but may introduce a *phase reversal.*

From the third enlargement images it is seen that the median filter completely eliminates very fine details while the mean filter blurs them but leaves something of the detail. This is related to the property of the mean filtering that it is an invertible process and, in principle, it is possible to deduce the original signal from the filtered signal. A typical characteristic of the median and many similar filters is that usually there are several signals that are mapped to a particular signal. Thus, it is impossible to deduce exactly the original signal from the filtered one.

In the smoother areas, like in the fourth enlargement images, the median performs better but introduces small squarelike artifacts. This phenomenon is characteristic of some nonlinear filters and is called *streaking.* The reason for the streaking becomes obvious if one considers a one-dimensional median filter. Suppose that we move the window one step ahead when one new sample enters the window and an old one leaves. If they both happen to be larger than the current median or smaller than the current median the output of the filter does not change. The probability of this event is 1/2 for a constant input contaminated by white noise. The probability that the output stays the same for longer segments is exponentially smaller but still high enough to cause streaking. In the case of two-dimensional signals, this effect is often called *blotching.*

Edge Preservation

By comparing the third enlargement images in Figure 1.16 and Figure 1.17 it is observed that the median filter preserves the edges much better than the mean filter. There is no blurring in the median filtered image while blurring is very visible in the corresponding image filtered by the mean filter. This blurring is clearly seen in the difference image. From the difference image, one can also see that the median filter causes some edge shifts.

1.3.4   Linear Versus Nonlinear Filtering

In the above examples nonlinear filtering (the median filter) performed much better than the corresponding linear filter (the mean filter). One should keep in mind that we were addressing a very specific filtering problem where

- the goal is to remove noise;

- the noise is impulsive;

- the signal and the noise occupy the same frequency band.

Nonlinear filtering has been very successful in solving these types of problems. It also performs very well in many image analysis tasks. Many morphological filtering methods, which are outside the scope of this book, form good examples of this field of image processing.

Because in nonlinear filtering we just drop the linearity requirement and by so doing enlarge the class of filters, nonlinear filtering, in principle, can never do worse than linear filtering. On the other hand, for the majority of signal processing tasks we do not know how to find a single nonlinear filter that would perform better than the optimal linear filter. However, except for rare ideal cases, it is certain that such nonlinear filters exist.

1.4   Algorithms

Let us now consider how the filters can be implemented by using some high-level programming language. Naturally, we are not only interested in finding a way to implement the filters but as well in knowing the computational complexity of the implementation. Even though the number of operations (addition, multiplication, function evaluation, and compare/swap) per one window position may not be high, when it is multiplied by the signal size it can be tremendous. The complexity of the filtering algorithm may severely hinder its usability and should be taken into consideration by any serious design engineer at an early stage of the filter design process. Naturally, many advanced techniques (parallel computers, dedicated hardware, etc.) may provide a remedy for a too long execution time. Our aim is not to study these techniques but just to provide basic implementation methods which anyone should be able to program in approximately one hour using his favorite programming language.

The mean filter is very straightforward to implement, (see Algorithm 1.1). In this algorithm 1 multiplication and N additions are needed, while no comparisons are required.

Median filtering is closely related to sorting and selection algorithms, which are classical topics in computer science. Several algorithms have been developed and their complexities have been evaluated. An interested reader may consult well-known books such as in References [21, 48]. The *quicksort* algorithm is often used in sorting. The number of compare/swap operations in the quicksort algorithm is $O(N \log N)$ on the average and $O(N^2)$ in the worst case. If the whole array is not needed to be ordered but only some specific order statistic should be found, like the median in our case, e.g., the *selection* algorithm (or its modifications) can be used [21]. Its average complexity is $O(N)$ but the worst case complexity is again $O(N^2)$. Algorithm 1.2 gives a general algorithm for the median filter.

Mean filter

Inputs: $NumberOfRows \times NumberOfColumns$ image
Moving window $W, |W| = N = 2k + 1$
Output: $NumberOfRows \times NumberOfColumns$ image

for $i = 1$ to $NumberOfRows$
    for $j = 1$ to $NumberOfColumns$
        place the window W at (i, j)
        let $Sum = 0$
        for every element X_m of the image inside W
            $Sum = Sum + X_m$
        end
        let $Output(i, j) = Sum/N$
    end
end

Algorithm 1.1. Algorithm for the mean filter.

Median filter

Inputs: $NumberOfRows \times NumberOfColumns$ image
Moving window $W, |W| = N = 2k + 1$
Output: $NumberOfRows \times NumberOfColumns$ image

for $i = 1$ to $NumberOfRows$
    for $j = 1$ to $NumberOfColumns$
        place the window W at (i, j)
        store the image values inside W in $\mathbf{x} = (X_1, X_2, \ldots, X_N)$
        find the median med of $\mathbf{x}$
        let $Output(i, j) = med$
    end
end

Algorithm 1.2. Algorithm for the median filter.

The mean and the median are running window filters. This can be used in the implementation by avoiding going through all the samples inside the window anew.[1] Consider first the one-dimensional mean filtering. When the window moves from position i to the next position $i + 1$, one new sample enters the window and one sample is left out. The other samples remain unchanged. Let X_o be the sample to be discarded, X_n to be the sample to be inserted, and $\overline{X_i}$ be the mean at the position i. The new mean is then easily found as

$$(1/N)(N\overline{X_i} - X_o + X_n).$$

Thus, we need two summations and two multiplications, one of which can be avoided using

---

[1] These ideas are so obvious that they have probably long been well known and used without mentioning in the early publications concerning these filters.

one extra memory cell. Naturally, as the mean filter is actually a convolution operation, it can also be efficiently implemented by using Fourier transforms.

A similar idea can also be used in running median filtering. Without going into details we present one way of calculating the running median. It can be carried out in the following steps [38]:

1. Set up the histogram of the pixels in the first window. Count the number of samples smaller than the median.

2. Move to the next window position. Update the histogram (requires modification of two histogram bins). Count the number *ltmdn* of samples which are smaller than the median of the previous step.

3. Start from the median of the previous step. Move up/down the histogram bins one at a time if the count *ltmdn* is smaller/greater than $(N + 1)/2$ and update *ltmdn* until the median is reached $(ltmdn = (N + 1)/2)$.

4. Stop if the whole signal is filtered. Otherwise go to the first step.

The complexity of this algorithm depends on how sparse the histogram will be and what is the difference between the values of two consecutive medians. Typically, the two consecutive medians are close to each other yielding a fast implementation.

If the number of possible values of the signal is much larger than the window size, as is the case, e.g., in floating point systems, histogram-based methods are useless. The running order statistics can be maintained by using the standard data structures that exist for priority queues [7].

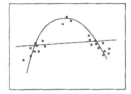

2

STATISTICAL PRELIMINARIES

YOU KNOW MY METHODS, WATSON
Arthur Conan Doyle, "Memoirs of Sherlock Holmes", 1893

TECHNIQUE IS NOTICED MOST MARKEDLY IN THE CASE OF THOSE
WHO HAVE NOT MASTERED IT.
Leon Trotsky, "Literature and Revolution, ch. 6", 1924

2.1 Random Variables and Distributions

2.1.1 One-Dimensional Random Variables

In signal and image processing the operations performed on the signals are of a deterministic nature and best expressed as algorithms that uniquely describe the relation between the input and the output signals. The signals and images, on the other hand, are best described in probabilistic terms in the framework of probability and statistics. One can say that the theory of random processes is the natural model for signals and images encountered in practice. An immediate consequence of this fact is that the methods developed in mathematical statistics are very useful in signal and image processing. In many signal processing tasks, e.g., in speech and audio signal processing, it is most natural to analyze and process the signals according to their frequency content, i.e., using the Fourier transform representations. In image processing where the frequency content of the signal has a less natural meaning than for electrical or audio signals, the strength of statistical methods is even more pronounced. A complicating factor is that even the basic statistical concepts are based on quite deep mathematical theories. This means that if we want to avoid simplifications that easily lead to false impressions we are forced to use fairly sophisticated tools from probability theory. Of course, one can gain an intuitive and useful understanding of the nonlinear filtering methods without going into the theoretical basics, but any serious analysis is then almost impossible.

In the following, we briefly review the statistical tools that are needed the most often in later sections. We do not intend to develop the concepts systematically; the main purpose is to fix the notation to be used in later sections. The readers who are not very familiar with the basic concepts of probability and statistics are encouraged to consult some of the many excellent books on the topic.

The key concept in applying statistics to signal processing is that measurements are modeled as realizations of *random variables*. Random variables are defined in a *probability space*.

Definition 2.1. A probability space is a triple $(\Omega, \mathcal{F}, P)$, where Ω is a set called the *sample space*, $\mathcal{F}$ is a σ-*algebra* on Ω and P is a *normed measure* defined on $\mathcal{F}$.

Remark 2.1. Recall that a collection of subsets of a set is a σ-algebra if it is closed under complementation and countable unions, and a function $P : \Omega \to [0, 1]$ is a normed measure on Ω if $P(A) \geq 0$ for all $A \in \mathcal{F}$, $P(\Omega) = 1$, and $P(A_1 \cup A_2 \cup \cdots) = P(A_1) + P(A_2) + \cdots$ whenever $A_1, A_2, \ldots$ are mutually disjoint elements of $\mathcal{F}$.

Example 2.1. Consider a throw of a die. The sample space is the set of the possible outcomes of the throw, i.e., $\{1, 2, 3, 4, 5, 6\}$, the faces of the die. The elements of the σ-algebra are the *events* we want to assign probabilities to:

$$A_0 = \emptyset,$$
$$A_1 = \{1\},$$
$$A_2 = \{2\},$$
$$\vdots$$
$$A_{63} = \{1, 2, 3, 4, 5, 6\}.$$

For each element A_i of $\mathcal{F}$ the probability $P(A_i)$ is defined in a consistent way so that the axioms of a probability space hold:

- $P(A_i) \geq 0$;

- $P(\Omega) = 1$;

- if $A_i \cap A_j = \emptyset$ then $P(A_i \cup A_j) = P(A_i) + P(A_j)$.

Remark 2.2. As long as the sample space is finite, everything is fairly straightforward. To be able to handle concrete filtering situations the sample space must be sufficiently complex, e.g., **R**. It is then not possible to simply take for $\mathcal{F}$ the family of all subsets because then it is not possible to assign probabilities to the members of $\mathcal{F}$ in a consistent and useful way. The solution is to take $\mathcal{F}$ to be the σ-algebra of all Borel sets of the real line, i.e., the smallest σ-algebra containing all countable unions of open intervals.

Remark 2.3. When we define the basic concepts, random variables, densities, etc., we emphasize the differences between different concepts by using the following notation. Random variables are denoted by capital Roman letters, e.g., X, Y, the realizations of random variables and constants are denoted by small case Roman letters, e.g. x, y, and the variables in functions and integration, etc., are denoted by small case Greek letters, e.g., ξ, η. Later, when discussing probability elements, estimation, etc., we drop this convention because it would no longer be natural and useful.

Definition 2.2. A *random variable*, defined in a probability space $(\Omega, \mathcal{F}, P)$, is a mapping $X : \Omega \to \mathbf{R}$. where $\mathbf{R}$ denotes the set of real numbers, such that

$$X^{-1}((-\infty, \xi]) = \{\omega \in \Omega : X(\omega) \le \xi\} \text{ is an element of } \mathcal{F}.$$

Definition 2.3. Let X be a random variable defined in the probability space $(\Omega, \mathcal{F}, P)$. The *distribution function* F_X of X is

$$F_X(\xi) = P(X(\omega) \le \xi) = P(X^{-1}((-\infty, \xi])).$$

The distribution function F_X of a random variable X has the following properties:

- F_X is nondecreasing;

- $\lim_{\xi \to -\infty} F_X(\xi) = 0$;

- $\lim_{\xi \to \infty} F_X(\xi) = 1$.

When we use random variables to model the observed values of a signal, the distribution function usually gives all the necessary information about the observation. The physical origin of the random variable, on the other hand, is reflected in the probability space. Sometimes it is necessary and helpful to consider also the underlying probability space to get more insight into the phenomenon.

If the distribution function of a random variable is differentiable, then the random variable models measurements that can have a continuous range of values. Such a random variable has a density function. Many characteristics of a random variable are easier to visualize from the density function than from the distribution function.

Definition 2.4. If F_X is differentiable the *probability density function* of the random variable X with the distribution function F_X is

$$f_X(\xi) = \frac{dF_X(\xi)}{d\xi}.$$

Remark 2.4. As the density function is obtained by differentiation from the distribution function, the distribution function can be obtained from the density by the following integration:

$$F_X(\xi) = \int_{-\infty}^{\xi} f_X(\eta) d\eta.$$

Example 2.2. Assume that the random phenomenon we are witnessing is the random selection of a point in a unit disk. Thus the sample space is

$$\Omega = \{(\omega_1, \omega_2) : \omega_1^2 + \omega_2^2 \le 1\},$$

and the natural model for the probability measure is to set

$$P(A) = \frac{1}{\pi} \times (\text{the area of } A),$$

where A is a subset of Ω. Assume that we can only measure the distance X of the selected point from the origin. The distance X is a random variable and the distribution function of X is

$$F_X(\xi) = P\{X \le \xi\} = \begin{cases} 0, & \text{if } \xi < 0, \\ \xi^2, & \text{if } 0 \le \xi < 1, \\ 1, & \text{if } 1 \le \xi. \end{cases}$$

Now, F_X is differentiable except at $\xi = 1$ and the probability density function of X is defined except at $\xi = 1$:

$$f_X(\xi) = \begin{cases} 0, & \text{if } \xi < 0, \\ 2\xi, & \text{if } 0 \le \xi < 1, \\ 0, & \text{if } 1 < \xi. \end{cases}$$

It is obvious that if a random variable X has a density function f_X defined on $a \le \xi \le b$, then

$$P(a \le \xi \le b) = \int_a^b f_X(\xi) d\xi.$$

We often encounter a random variable that can only attain a finite number of values. Let the random variable X attain value $\xi_i \in \mathbf{R}$ with probability $p(\xi_i)$, $i = 1, 2, \ldots, n$, where $\sum_{i=1}^n p(\xi_i) = 1$. We say that X has a *discrete distribution*, and obviously its distribution function is

$$F_X(\xi) = \sum_{\xi_i \le \xi} p(\xi_i).$$

The actual physical phenomenon that is generating the signal to be processed is usually a combination of many factors, some giving rise to continuous behavior and some to a more discrete nature. If a random variable has both discrete and continuous characteristics then it has a distribution function with strictly increasing continuous segments as well as steplike discontinuities.

Example 2.3. Assume that a random variable takes the value $1/2$ with probability $1/3$ and that any other value in the interval $[0,1]$ is equally likely. The distribution function is now

$$F_X(\xi) = \begin{cases} 0, & \text{if } \xi < 0, \\ \frac{2\xi}{3}, & \text{if } 0 \le \xi < \frac{1}{2}, \\ \frac{1}{3} + \frac{2\xi}{3}, & \text{if } \frac{1}{2} \le \xi < 1, \\ 1, & \text{if } 1 \le \xi. \end{cases} \tag{2.1}$$

Many operations become simpler if we can also express those random variables with steplike discontinuities in their distribution functions using probability density functions. The key idea is to let a density function contain "impulses" that serve as coding for the locations and the magnitudes of the jumps in the distribution function. We define the *δ-function* by the following conditions:

- $\delta(\xi) = 0$ for $\xi \ne 0$;

- $\int_{-\infty}^\infty \delta(\xi) d\xi = 1$.

Note that we have $\int_{-\infty}^{\infty} g(\xi)\delta(\eta_0 - \xi)d\xi = g(\eta_0)$ for all g, and so the δ-function can be used to pick values of a function by the integration operation.

Example 2.4. Consider the random variable defined in the previous example. Its distribution function is given in (2.1). If we write

$$g(\xi) = \begin{cases} \frac{2}{3}, & \text{if } 0 \leq \xi \leq 1, \xi \neq \frac{1}{2}, \\ 0, & \text{otherwise,} \end{cases}$$

then the density of X can be written as

$$f_X(\xi) = g(\xi) + \frac{1}{3}\delta(\xi - \frac{1}{2}). \tag{2.2}$$

The expression (2.2) does not define a function in the ordinary sense. We usually do not need the actual values of probability densities but the results when they are manipulated with operations involving integrals. Therefore, this does not cause any harm as long as one keeps in mind the special nature of the δ-function.

The distribution function completely describes the behavior of a random variable, but it is difficult to visualize the properties of a random variable or compare their properties using distribution functions. We need more compact numerical characteristics of random variables. Such characteristics can in many cases be expressed as *expectations*.

Definition 2.5. Let X be a random variable with the probability density function $f_X(\xi)$. The expectation of X is

$$E\{X\} = \int_{-\infty}^{\infty} \xi f_X(\xi)d\xi.$$

The intuitive interpretation of the expectation is that if we consider many realizations of a random variable then the average value of these realizations should be close to the expectation.

Let X be a random variable and $g : \mathbf{R} \rightarrow \mathbf{R}$ a function. Then the relation

$$Y = g(X)$$

defines a new random variable. Its distribution function could be written out as

$$F_Y(\xi) = P\{Y \leq \xi\} = P(X^{-1}(g^{-1}((-\infty, \xi]))),$$

but usually it is not necessary to work with this expression because one can show that the following very important relation holds:

$$E\{Y\} = \int_{-\infty}^{\infty} g(\xi)f_X(\xi)d\xi.$$

The most important numerical characteristics of a random variable are its *moments*

$$\mu_k' = E\{X^k\}$$

and its *central moments*

$$\mu_k = E\{(X - \mu_1')^k\}.$$

The first moment is the *expectation* (or *mean*) of a random variable and it is denoted by μ. The second central moment μ_2 is the *variance* of the random variable and denoted by σ^2 or $\mathrm{Var}\{X\}$. The intuitive meaning of the variance is that if we consider many realizations of a random variable then it describes how widely spread are the values of the realizations.

Even though the expectation and the variance tell little about a general random variable, in practice, we can often assume that its distribution function has a certain form and then, knowing its expectation and variance may specify it completely.

Example 2.5. The most important distribution is the *normal* or *Gaussian* distribution $N(\mu, \sigma^2)$ defined by its density function

$$f_N(\xi) = \frac{1}{\sqrt{2\pi}\sigma} e^{-\frac{(\xi-\mu)^2}{2\sigma^2}}.$$

It is completely specified once the expectation (mean) μ and the variance σ^2 are given. Notice that large positive or negative values have very small probabilities. In fact for $\mu = 0$ and $\sigma^2 = 1$ we have for large ξ_0 (Exercise)

$$P\{X > \xi_0\} = \int_{\xi_0}^{\infty} f_N(\xi)d\xi \approx \frac{1}{\sqrt{2\pi}\xi_0} e^{-\frac{\xi_0^2}{2}}.$$

Example 2.6. A distribution that better models a random phenomenon where also large positive and negative values are probable is the *Laplace* or *biexponential* distribution defined by

$$f_L(\xi) = \frac{\alpha}{2} e^{-\alpha|\xi-\mu|}, \quad \alpha > 0.$$

It is also completely specified by the expectation μ and the variance $\sigma^2 = 2\alpha^{-2}$. The probability of large values is much higher than for the normal distribution. If $\mu = 0$ and $\sigma^2 = 1$ we have

$$P\{X > \xi_0\} = \int_{\xi_0}^{\infty} f_L(\xi)d\xi = \frac{1}{2} e^{-\sqrt{2}\xi_0}.$$

Example 2.7. A distribution where large positive or negative values are very likely is the *Cauchy distribution* defined by its density

$$f_C(\xi) = \frac{1}{\pi} \frac{\sigma}{\sigma^2 + (\xi - \mu)^2}. \tag{2.3}$$

It is also completely specified if the parameters μ and σ are known. The parameter μ gives the location of the distribution and the parameter σ indicates scale, i.e., the broadness of the distribution. However, a random variable with a Cauchy distribution does not even have the expectation, not to speak of the variance, because

$$\int_{-\infty}^{\infty} \xi f_C(\xi)d\xi$$

does not exist as an improper integral (Exercise). For the Cauchy distribution with $\mu = 0$ and $\sigma^2 = 1$ we have for large ξ_0

$$P\{X > \xi_0\} = \int_{\xi_0}^{\infty} f_C(\xi)d\xi \approx \frac{1}{\pi\xi_0}, \tag{2.4}$$

and so large positive and negative values occur very often.

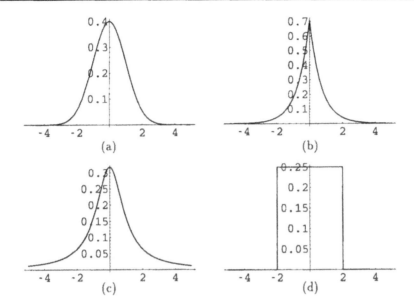

Figure 2.1. Density functions of (a) normal distribution, $\mu = 0$, $\sigma^2 = 1$; (b) Laplace distribution, $\mu = 0$, $\sigma^2 = 1$; (c) Cauchy distribution, $\mu = 0$, $\sigma^2 = 1$; (d) uniform distribution, $x_0 = -2$, $x_1 = 2$.

Example 2.8. A frequently used distribution is the *uniform* (or *rectangular*) distribution defined by its density function

$$f_U(\xi) = 1/(x_1 - x_0), \quad x_0 \le \xi \le x_1.$$

Here, the large positive and negative values are not possible at all, and the distribution can be said to have no tails.

The densities of these four important distributions are shown in Figure 2.1. It is clearly seen that the uniform distribution does not have tails, the normal distribution has the next weakest tails, and the Cauchy distribution has the strongest tails. The strength of the tails is directly related to the occurrence of very large positive and negative values.

Two other important numerical characteristics of random variables are the *median* and the *mode*.

Definition 2.6. Let X be a random variable whose distribution function is F_X. The point x_{med} is the median of X if

$$F_X(x_{\text{med}}) = 1/2.$$

Notice that the median does not necessarily exist and that it may be nonunique.

Definition 2.7. Let X be a random variable with a continuous probability density function $f_X(\xi)$. A point x_{mode} is the mode of X if

$$f_X(x_{\text{mode}}) \ge f_X(\xi) \quad \text{for all } \xi.$$

Notice that the mode does not necessarily exist and that it may be nonunique.

Example 2.9. The mode and the median of the normal distribution $N(\mu, \sigma^2)$, the Laplace distribution and the Cauchy distribution are equal to μ. The median of the random variable in (2.1) does not exist (Exercise).

2.1.2   Random Vectors and Random Processes

We usually need to manipulate several random variables simultaneously. For instance, all the pixels of an image can be thought of as realizations of random variables. To be able to model the interplay of the different components or pixels we need the concept of a *multidimensional random variable*, also called a *random vector* or a *finite dimensional stochastic process*.

Let us consider a probability space $(\Omega, \mathcal{F}, P)$ on which N random variables $X_1, X_2, \ldots, X_N$ are defined. As for all $i = 1, 2, \ldots, N$, $X_i^{-1}((-\infty, \xi]) \in \mathcal{F}$ for any ξ, the fact that $\mathcal{F}$ is a σ-algebra guarantees that

$$S = \{\omega \in \Omega : X_1(\omega) \leq \xi_1, X_2(\omega) \leq \xi_2, \ldots, X_N(\omega) \leq \xi_N\}$$

is an element of $\mathcal{F}$, and thus $P(S)$ is defined. This means that we have a well defined *joint distribution function* of random variables on the same probability space.

Definition 2.8. Let $X_1, X_2, \ldots, X_N$ be random variables defined on a probability space $(\Omega, \mathcal{F}, P)$. Their joint distribution function is

$$F(\xi_1, \xi_2, \ldots, \xi_N) = P\{X_1 \leq \xi_1, X_2 \leq \xi_2, \ldots, X_N \leq \xi_N\}.$$

If the N-fold mixed derivative of F exists we can define *the joint probability density function* of the random vector $(X_1, X_2, \ldots, X_N)$ similar to the one-dimensional case. Equivalently, we can define as the following:

Definition 2.9. Let $X_1, X_2, \ldots, X_N$ be random variables defined on $(\Omega, \mathcal{F}, P)$ and their joint distribution function be $F(\xi_1, \ldots, \xi_N)$. If there exists a function $f(\eta_1, \eta_2, \ldots, \eta_N)$ such that

$$F(\xi_1, \xi_2, \ldots, \xi_N) = \int_{-\infty}^{\xi_1} \cdots \int_{-\infty}^{\xi_N} f(\eta_1, \eta_2, \ldots, \eta_N) d\eta_1 d\eta_2 \cdots d\eta_N,$$

then $f(\eta_1, \eta_2, \ldots, \eta_N)$ is the joint probability density function of $X_1, X_2, \ldots, X_N$.

The probability density has the property that, for any $G \subset \mathbf{R}^N$, we have

$$P\{(X_1, X_2, \ldots, X_N) \in G\} = \int \cdots \int f(\eta_1, \eta_2, \ldots, \eta_N) d\eta_1 d\eta_2 \cdots d\eta_N,$$

where the integration is over G.

The density of any of the random variables, often called the marginal density, is obtained by "integrating out" the rest of the variables:

$$f_{X_1}(\xi_1) = \int_{-\infty}^{\infty} \cdots \int_{-\infty}^{\infty} f(\xi_1, \eta_2, \eta_3, \ldots, \eta_N) d\eta_2 \eta_3 \cdots d\eta_N,$$

$$f_{X_2}(\xi_2) = \int_{-\infty}^{\infty} \cdots \int_{-\infty}^{\infty} f(\eta_1, \xi_2, \eta_3 \ldots, \eta_N) d\eta_1 d\eta_3 \cdots d\eta_N,$$

$$f_{X_N}(\xi_N) = \int_{-\infty}^{\infty} \cdots \int_{-\infty}^{\infty} f(\eta_1, \eta_2, \ldots, \eta_{N-1}, \xi_N) d\eta_1 \cdots d\eta_{N-1}.$$

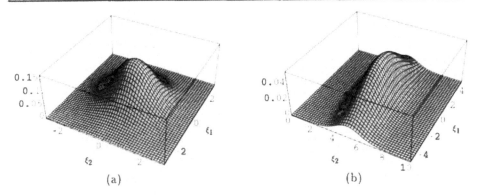

Figure 2.2. Density functions of two-dimensional normal distributions (a) $\mu_1 = \mu_2 = 0$, $\sigma_1^2 = \sigma_2^2 = 1$, $\rho = 0$; (b) $\mu_1 = 0$, $\mu_2 = 5$, $\sigma_1^2 = 3$, $\sigma_2^2 = 1$, $\rho = 0$.

Example 2.10. The density of a *two-dimensional normal distribution* is

$$f(\xi_1, \xi_2) =$$
$$= \frac{1}{2\pi\sigma_1\sigma_2\sqrt{1-\rho^2}} \exp\left\{ -\frac{1}{2(1-\rho^2)} \left(\frac{(\xi_1 - \mu_1)^2}{\sigma_1^2} - 2\rho\frac{(\xi_1 - \mu_1)(\xi_2 - \mu_2)}{\sigma_1\sigma_2} + \frac{(\xi_2 - \mu_2)^2}{\sigma_2^2} \right) \right\},$$

where μ_1 and μ_2 are the expectations of X_1 and X_2, σ_1^2 and σ_2^2 are the variances of X_1 and X_2 (Exercise), and ρ is the correlation coefficient of X_1 and X_2 (See Definition 2.11 for the definition). Figure 2.2 shows some selected two-dimensional normal distributions.

Like one-dimensional random variables, random vectors have analogous numerical parameters that characterize their properties. The most important ones are again defined by expectations.

Definition 2.10. Let $(X_1, X_2, \ldots, X_N)$ be a random vector with joint density function $f(\xi_1, \xi_2, \ldots, \xi_N)$ and $g : \mathbf{R}^N \to \mathbf{R}$ a function. The *expectation of the random variable*

$$Y = g(X_1, X_2, \ldots, X_N)$$

is defined as

$$E\{g(X_1, X_2, \ldots, X_N)\} = \int_{-\infty} \cdots \int_{-\infty}^{\infty} g(\xi_1, \xi_2, \ldots, \xi_N) f(\xi_1, \xi_2, \ldots, \xi_N) d\xi_1 d\xi_2 \cdots d\xi_N.$$

Remark 2.5. The expectation is a linear operation on the random variables, i.e.,

$$E\{\alpha g(X_1, X_2, \ldots, X_N) + \beta h(X_1, X_2, \ldots, X_N)\}$$
$$= \alpha E\{g(X_1, X_2, \ldots, X_N)\} + \beta E\{h(X_1, X_2, \ldots, X_N)\}$$

for all functions g, h, and complex numbers α, β.

Definition 2.11. The *mean* of the random variable X_i, $i = 1, 2, \ldots, N$, is

$$\mu_i = E\{X_i\} = \int_{-} \cdots \int_{-\infty}^{\infty} \xi_i f(\xi_1, \xi_2, \ldots, \xi_N) d\xi_1 d\xi_2 \cdots d\xi_N$$

and the *covariance* of X_i and X_j is

$$c_{ij} = E\{(X_i - \mu_i)(X_j - \mu_j)\} = \int_{-\infty}^{\infty} \cdots \int_{-\infty}^{\infty} (\xi_i - \mu_i)(\xi_j - \mu_j) f(\xi_1, \xi_2, \ldots, \xi_N) d\xi_1 d\xi_2 \cdots d\xi_N.$$

The $N \times N$ matrix $\mathbf{C} = [c_{ij}]$ is called the *covariance matrix* of the random variables $X_1, X_2, \ldots, X_N$. It is always positive semidefinite, i.e., $\mathbf{x}^T \mathbf{C} \mathbf{x} \geq 0$ for all $\mathbf{x}$. The element c_{ij} of $\mathbf{C}$ indicates the interrelation between X_i and X_j. When normalized, it is called the *correlation coefficient* $r_{ij} = c_{ij}/(\sqrt{c_{ii}c_{jj}})$. Clearly, by definition $c_{ij} = c_{ji}$ and $r_{ij} = r_{ji}$ for all i, j.

Example 2.11. *Multidimensional normal distribution.* The probability density function of an N-dimensional normally distributed random vector $(X_1, X_2, \ldots, X_N)$ is

$$f(\xi_1, \xi_2, \ldots, \xi_N) = \frac{1}{(2\pi)^{N/2}|\mathbf{C}|^{1/2}} e^{-\frac{1}{2}(\boldsymbol{\xi} - \boldsymbol{\mu})^T \mathbf{C}^{-1}(\boldsymbol{\xi} - \boldsymbol{\mu})} \tag{2.5}$$

where $\boldsymbol{\xi} = (\xi_1, \xi_2, \ldots, \xi_N)^T$, $\boldsymbol{\mu} = (\mu_1, \mu_2, \ldots, \mu_N)^T$, $\mu_i = E\{X_i\}$, $i = 1, 2, \ldots, N$, $\boldsymbol{\mu}^T$ denotes the transpose of $\boldsymbol{\mu}$,

$$\mathbf{C} = [c_{ij}] = [E\{(X_i - \mu_i)(X_j - \mu_j)\}],$$

$\mathbf{C}^{-1}$ denotes the inverse of the matrix $\mathbf{C}$, and $|\mathbf{C}|$ denotes the determinant of the matrix $\mathbf{C}$. It is a straightforward exercise to show that Example 2.10 can be obtained as a special case.

Recall that two events $A, B \in \mathcal{F}$ of the probability space $(\Omega, \mathcal{F}, P)$ are *independent* if $P\{A \cap B\} = P\{A\}P\{B\}$. The random variables $X_1, X_2, \ldots, X_N$ are *independent* if the joint distribution function $F(\xi_1, \xi_2, \ldots, \xi_N)$ satisfies

$$F(\xi_1, \xi_2, \ldots, \xi_N) = F(\xi_1)F(\xi_2) \cdots F(\xi_N).$$

Then the joint density satisfies

$$f(\xi_1, \xi_2, \ldots, \xi_N) = f(\xi_1)f(\xi_2) \cdots f(\xi_N).$$

Example 2.12. Consider the multivariate normal distribution (2.5). From linear algebra we know that there is an orthogonal matrix, $\mathbf{Q}$ such that

$$\mathbf{Q}^T \mathbf{C} \mathbf{Q} = \lceil \sigma_1^2, \sigma_2^2, \ldots, \sigma_N^2 \rfloor,$$

that is, a diagonal matrix with $\sigma_1^2, \sigma_2^2, \ldots, \sigma_N^2$ on the diagonal. Recall that a matrix $\mathbf{Q}$ is orthogonal if $\mathbf{Q}^T \mathbf{Q} = \mathbf{Q}\mathbf{Q}^T = I$. If we make in (2.5) the substitution (or change of variables)

$$\boldsymbol{\eta} = \mathbf{Q}^T \boldsymbol{\xi}, \quad \boldsymbol{\xi} = \mathbf{Q}\boldsymbol{\eta}$$

and denote $\mathbf{m} = \mathbf{Q}^T \boldsymbol{\mu}$ we obtain

$$\frac{1}{2}(\boldsymbol{\xi} - \boldsymbol{\mu})^T \mathbf{C}^{-1}(\boldsymbol{\xi} - \boldsymbol{\mu}) = \frac{1}{2}(\boldsymbol{\eta} - \mathbf{m})^T (\mathbf{Q}^T \mathbf{C} \mathbf{Q})^{-1}(\boldsymbol{\eta} - \mathbf{m}) = \sum_{i=1}^{N} \frac{(\eta_i - m_i)^2}{2\sigma_i^2}.$$

This means that if we have an N-dimensional normal distribution it is always possible to make a linear transformation of the random variables that makes the correlation matrix diagonal. We can, of course, perform this "decorrelation" to any random vector, but only for the normal distribution does this decorrelation result in independent random variables.

Recall that when A and B are events of $(\Omega, \mathcal{F}, P)$ the *conditional probability of A assuming B* is

$$P(A|B) = \frac{P(A \cap B)}{P(B)} \tag{2.6}$$

provided that $P(B) \neq 0$. If X and Y are random variables with continuous distributions the *conditional density of X assuming $Y = y$* is

$$f(\xi|y) = \frac{f_{XY}(\xi, y)}{f_Y(y)}.$$

The resulting random variable is called the conditional random variable X assuming $Y = y$. For each y there is then the *conditional expectation*

$$E\{X|y\} = \int_{-\infty}^{\infty} \xi f(\xi|y)d\xi \tag{2.7}$$

if the integral (2.7) exists.

The conditional expectation $E\{X|y\}$ is a function of y. An important fact is that this function is the solution of the following nonlinear mean square estimation problem.

Example 2.13. Consider realizations (x, y) of a random vector (X, Y) with density function $f(\xi, \eta)$. Assume that we can only observe y but have no other information about the corresponding x than that provided by the density function. We need to find an estimator for x in the form $T(y)$ such that the mean square error

$$E\{(X - T(Y))^2\} = \int_{-\infty}^{\infty} (\xi - T(\eta))^2 f(\xi, \eta) d\xi d\eta \tag{2.8}$$

is minimized. It can be shown (Exercise) that the solution is

$$T(y) = E\{X|y\}. \tag{2.9}$$

This is called the *regression line*.

2.1.3 Probability Elements

Consider a random variable X that has the density function $f(x)$. It is sometimes convenient to use the ordinary *differential notation* and write

$$P(x_0 < x < x_0 + dx) = f(x_0)dx$$

understanding, of course, that one must be careful as always when dealing with differentials. The quantity $f(x_0)dx$ is called the *probability element*. Occasionally, it is simpler to find the probability element of a random variable than directly determine the distribution function. The probability element gives the density, and integration then gives the distribution function.

Example 2.14. Assume that X is uniformly distributed on the interval $[-\pi/2, \pi/2]$ and $Y = \tan(X)$. Then equating the probability elements gives

$$f_Y(y)dy = \frac{1}{\pi}dx$$

or

$$f_Y(y) = \frac{1}{\pi}\frac{dx}{dy} = \frac{1}{\pi}\frac{d}{dy}\arctan(y) = \frac{1}{\pi}\frac{1}{1 + y^2},$$

so that Y has the Cauchy distribution characterized by its density given in (2.3).

2.1.4   Transformations of Random Vectors

Consider the one-dimensional function $g : [a, b] \rightarrow [c, d]$, which for simplicity we assume to have a nonvanishing continuous derivative on (a, b). Let X be a random variable defined on $[a, b]$ where it has the density $f_X(x)$. Then the density function of $Y = g(X)$ on $[c, d]$ is given by

$$f_Y(y) = f_X(g^{-1}(y)) \left| \frac{dg^{-1}(y)}{dy} \right|. \tag{2.10}$$

This means that the transformation $y = g(x)$ "carries" the probability element $f_X(x)dx$ to $f_Y(y)dy$, where $f_Y(y)dy$ is computed from (2.10). The formula (2.10) is exactly the same as the familiar rule for changing the variable of integration, and this is obvious because the probabilities are calculated by integration from the density function.

For higher dimensional random vectors the principle stays the same but the role of $|dx/dy|$ is played by the *Jacobian of the transformation*. The exact conditions become more complicated, but often it is enough to apply the simple formula for transforming the probability element for a one-to-one transformation and, if necessary, verify the final result by other means.

Let $\mathbf{x} = (X_1, X_2, \ldots, X_N)$ be a random vector with the density $f_{\mathbf{X}}(x_1, x_2, \ldots, x_N)$ and $\mathbf{y} = (y_1, y_2, \ldots, y_N)^T = \mathbf{g}(\mathbf{x}) = (g_1(x_1, x_2, \ldots, x_N), \ldots, g_N(x_1, x_2, \ldots, x_N))^T$ a differentiable one-to-one transformation with the inverse $\mathbf{g}^{-1}$. Then the density function of $\mathbf{y} = \mathbf{g}(\mathbf{x})$ is given by

$$f_{\mathbf{Y}}(\mathbf{y}) = f_{\mathbf{X}}(\mathbf{g}^{-1}(\mathbf{y})) \left| \frac{\partial(x_1, x_2, \ldots, x_N)}{\partial(y_1, y_2, \ldots, y_N)} \right|,$$

where $\partial(x_1, x_2, \ldots, x_N)/\partial(y_1, y_2, \ldots, y_N)$ is the Jacobian of the inverse transform:

$$\begin{cases} x_1 = (\mathbf{g}^{-1})_1(y_1, y_2, \ldots, y_N), \\ \quad \vdots \\ x_N = (\mathbf{g}^{-1})_N(y_1, y_2, \ldots, y_N), \end{cases}$$

that is, the determinant

$$\frac{\partial(x_1, x_2, \ldots, x_N)}{\partial(y_1, y_2, \ldots, y_N)} = \left| \frac{\partial x_i}{\partial y_j} \right|.$$

Thus the probability element

$$f_{\mathbf{X}}(x_1, x_2, \ldots, x_N)dx_1 dx_2 \cdots dx_N$$

is transformed to the probability element

$$f_{\mathbf{Y}}(y_1, y_2, \ldots, y_N)dy_1 dy_2 \cdots dy_N = f_{\mathbf{X}}(\mathbf{g}^{-1}(\mathbf{y})) \left| \frac{\partial(x_1, x_2, \ldots, x_N)}{\partial(y_1, y_2, \ldots, y_N)} \right| dy_1 dy_2 \cdots dy_N. \tag{2.11}$$

Remark 2.6. If the transformation $\mathbf{y} = \mathbf{g}(\mathbf{x})$ is not one-to-one, then it is necessary to add the probability elements corresponding to all solutions $\mathbf{x}$ of $\mathbf{y} = \mathbf{g}(\mathbf{x})$.

Example 2.15. Let the random vector $\mathbf{x} = (X, Y)$ have the joint density

$$f_{\mathbf{X}} = \begin{cases} 1, & \text{for } 0 \leq x \leq 1, \ \ 0 \leq y \leq 1 , \\ 0, & \text{otherwise,} \end{cases}$$

and the random vector (U, V) be defined by the following transformation:

$$\begin{cases} u = \sqrt{-2 \ln x} \cos(2\pi y), \\ v = \sqrt{-2 \ln x} \sin(2\pi y). \end{cases}$$

Now, the inverse transform is given by

$$\begin{cases} x = e^{-\frac{u^2+v^2}{2}}, \\ y = \frac{1}{2\pi} \arctan(v/u), \end{cases}$$

and so the Jacobian becomes

$$\left| \frac{\partial(u, v)}{\partial(x, y)} \right| = \frac{1}{2\pi} e^{-\frac{u^2+v^2}{2}}.$$

Thus, we see that because the probability element of (X, Y) is $1 \cdot dxdy$, the probability element of (U, V) is $\frac{1}{2\pi} e^{-\frac{u^2+v^2}{2}} dudv$, that is, the probability element of a circularly symmetric Gaussian random vector, where the correlation coefficient ρ equals 0, and for the variances it holds that $\sigma_1^2 = \sigma_2^2$.

The above relation gives a useful way to generate normal random variables because it is usually easy to generate random variables that are uniformly distributed on $[0, 1]$ (Exercise).

2.2 Signal and Noise Models

This book will mostly concentrate on nonlinear filtering techniques for removing an unknown corruption from a signal. In Chapter 1, we briefly discussed the modeling of the original signal and the process that corrupts it. The general model consists of the random process (or vector) $s = (S_1, S_2, \ldots, S_n)$ representing the original signal, the random process $x = (X_1, X_2, \ldots, X_n)$ representing the corrupted signal, together with the joint density $f_{S,X}(s, x)$ that models the corruption mechanism.

The simplest and most important case is the case of *additive white noise* meaning that

$$(X_1, X_2, \ldots, X_n) = (S_1, S_2, \ldots, S_n) + (N_1, N_2, \ldots, N_n),$$

where the concept "additive" contains also the assumption that N_i are independent of S_i and the "whiteness" contains the assumption that the N_is are mutually independent (or at least uncorrelated). The importance of this case stems from the facts that these assumptions are often quite realistic, e.g., in communication theory, and under these assumptions it is possible to explicitly design optimal linear filters. This is particularly true if we assume that the N_is are independent and identically distributed normal random variables, in other words, additive white Gaussian noise.

In image processing the assumption of additive white noise seldom holds. The intensity of an image formed by an image acquisition system is usually multiplicative with respect to illumination and the reflectivity of the observed surface. This as well as many remote sensing systems leads to *multiplicative noise* [43] modeled as

$$(X_1, X_2, \ldots, X_n) = (S_1 N_1, S_2 N_2, \ldots, S_n N_n),$$

where again the N_is are usually taken as independent of the S_is.

In many image processing applications, the noise processes are far more complex, and the dependencies between the noise and the signal are very complicated. In Chapter 1, we saw that impulsive noise causes difficulties in signal processing. In particular, linear methods largely fail for impulsive noise. How should we define impulsive noise and how could it be modeled? We say that a noise process is impulsive if as a result many of the signal values do not change at all or change slightly and some signal values change "dramatically", in other words, the change is clearly visible.

When we are considering realistic noise models for image processing there are some points that are often overlooked. Perhaps the most important is to remember that in practice the same number of bits will be used to represent the noisy and the noise-free signal. Notice that if we use the fixed point quantization usual in image processing, then the signal is quantized, e.g., to the 256 levels $0, 1, \ldots, 255$. As one wants to get the best gray scale resolution with the fixed number of gray scale levels, the scaling is chosen so that the signal occupies the whole range. This means that the additive noise will be clipped to fit the range, and we cannot have additive noise that is independent of the signal even if the original noise source were of that nature. On the other hand, if one uses floating point representation, where the dynamic range is so large that virtually no values are clipped, then the additive model is more realistic.

There are many models for impulsive noise. For example, the impulses may have different amplitude values, or all the nonzero values are the same. Common for the models of impulsive noise in images is the appearance of noise as black and/or white spots in images, i.e., the noisy pixels have either a very small or a very large value. This type of noise is often called salt-and-pepper noise because one could create it by sprinkling salt-and-pepper on an image. Pure salt-and-pepper noise is very easy to remove from images because the maximal values occur rarely in actual images and thus just checking whether the pixel has a maximal or minimal value reveals if it is corrupted or not. A more realistic impulsive noise is modeled as bit errors in the signal values. Typical sources for this kind of noise are channel errors in communication or storage.

Impulses are also referred to as *outliers*. In statistics, outliers can be defined as observations which appear to be inconsistent with the remainder of the data [12]. Another definition for outliers includes a basic assumption of the nature of the data and states that "an outlier is an observation which appears suspicious in the light of some provisional initial assignment of a probability model to explain the data generating process" [12].

The following impulsive noise models have been proposed in the literature cf. (e.g., Reference [44]):

Model 1 (One-sided fixed impulses): At every signal component (time instant or pixel) an error occurs with probability p independent of both the errors at other signal points and the values of the original signal. An erroneous point has the fixed value d. Other points remain unaltered. Let $s(i)$ denote the original signal and $x(i)$ denote the distorted signal, in which case

$$x(i) = \begin{cases} d, & \text{with probability } p, \\ s(i), & \text{with probability } 1 - p. \end{cases}$$

It is important to notice that this model cannot be expressed as additive noise that is independent of the signal.

Model 2 (Two-sided fixed impulses): At every signal component an error occurs with probability $p + q$ independent of both the errors at other signal points and the values

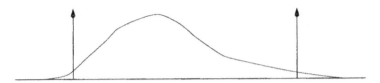

Figure 2.3. The density of a mixed-type noise distribution. The impulses are denoted by arrows pointing upwards.

of the original signal. An erroneous point has the fixed value c with probability $\frac{p}{p+q}$; otherwise it receives the fixed value d. Other points remain unaltered. Let $s(i)$ be the original signal and $x(i)$ be the distorted signal. Then, the model can be expressed in the following way:

$$x(i) = \begin{cases} c, & \text{with probability } p, \\ d, & \text{with probability } q, \\ s(i), & \text{with probability } 1 - p - q. \end{cases}$$

This kind of noise cannot be expressed as additive noise that is independent of the signal.

Model 3 (Mixed impulsive noise): One often encounters a situation where noise comes from two or more separate sources. For instance, there is thermal noise before A/D conversion that is well modeled as white Gaussian noise. There may also be faulty components in the digital circuitry resulting, e.g., in bit changes in the digitized signal. The resulting overall noise could be modeled, though not exactly, as additive noise where the noise distribution has both continuous and discrete characteristics.

Let $s(i)$ be the original signal values. The corrupted signal values then have the form

$$x(i) = s(i) + n(i),$$

where $n(i)$ are *independent and identically distributed (i.i.d.)* random variables having the same density

$$g(u) = (1 - p)f(u) + \frac{p}{2}(\delta(d - u) + \delta(c - u)).$$

Intuitively this means that the additive noise in each component is such that with probability $(1 - p)$ it appears like the noise following the continuous density $f(u)$ and with probability $p/2$ it has either the value d or c. This kind of mixture density is illustrated in Figure 2.3, where $f(u)$ is a Gaussian density.

Model 4 (Bit errors): Let

$$s(i) = k_1(i)2^{B-1} + k_2(i)2^{B-2} + \cdots + k_{B-1}(i)2 + k_B(i),$$

where $k_j(i) \in \{0, 1\}$ for all i, j, be the original signal values quantized to B bits. Assume that the bit errors occur with probability p independent both of the errors at other signal values and errors in this signal value. Then the corrupted signal values are of the form

$$x(i) = k_1^*(i)2^{B-1} + k_2^*(i)2^{B-2} + \cdots + k_{B-1}^*(i)2 + k_B^*(i),$$

where

$$k_j^*(i) = \begin{cases} k_j(i), & \text{with probability } 1 - p, \\ 1 - k_j(i), & \text{with probability } p. \end{cases}$$

Naturally, this model cannot be expressed as additive noise that is independent of the signal.

2.3   Estimation

2.3.1   Point Estimation

This section briefly studies *point estimation* and its relation to filtering, especially to the mean and the median filters. It should be emphasized that here we only address some particular topics in estimation theory. Our intention is to recall those basic principles of estimation that are closely connected with nonlinear signal processing and, in particular, image processing. As we have already discussed in Chapter 1, filtering noise from a signal is nothing but estimating the underlying "true" signal value. An immediate consequence is that many point estimators can be interpreted as filters. We shall discuss this connection more extensively in Chapter 3 where many useful filters that are based on estimation theory are introduced.

Suppose, for example, that we have a population that is normally distributed with variance σ^2 and we want to estimate the mean μ of the distribution. Let a sample $X_1, X_2, \ldots, X_N$ be taken of the population. Mathematically, this means that we observe N independent random variables $X_1, X_2, \ldots, X_N$ having a normal distribution $N(\mu, \sigma^2)$. The task in point estimation is to pick a statistic (i.e., a function) $T(X_1, X_2, \ldots, X_N)$ that will form an estimate for μ from $X_1, X_2, \ldots, X_N$ in some reasonable way. The numerical value of $T(\cdot)$ is called an *estimate* of μ, while the statistic T is called an *estimator* of μ.

Let X be a random variable in a probability space $(\Omega, \mathcal{F}, P)$. Suppose that the distribution function $F(\cdot)$ of X depends on certain parameters $\theta_1, \theta_2, \ldots, \theta_k$. Let $\theta = (\theta_1, \theta_2, \ldots, \theta_k)$ be the vector of the parameters associated with $F(\cdot)$. The set of all possible values of the parameters of the distribution function $F(\cdot)$ is called the *parameter set*. We use the notation $F(\cdot; \theta)$ $(f(\cdot; \theta))$ for the distribution (density) function of X if θ is the parameter vector. The parameter set is denoted by Θ.

Definition 2.12. Let $X_1, X_2, \ldots, X_N$ be a random sample from $F(\cdot; \theta)$, where $\theta \in \Theta \subseteq \mathbf{R}$. In other words, the length of the parameter vector is one. To emphasize this, we will denote the real parameter by θ. A statistic $T(X_1, X_2, \ldots, X_N)$ is said to be a point estimator of θ if $T(\cdot)$ maps $\mathbf{R}^n$ onto Θ.

The idea of parametric point estimation is to find an estimator $T(\cdot)$ that optimally distills information about the unknown parameter θ from the random sample.

Remark 2.7. Clearly, the mean filter and the median filters are estimators distilling information about the unknown parameter—the original value of the component of the signal that we are filtering and that is corrupted by noise. The signal values inside the current window form the random sample. To distinguish the mean and the median from those of a distribution, the mean and the median taken from a sample are called the *sample mean* and the *sample median*.

In the sequel, we will use $X_{(1)} \leq X_{(2)} \leq \cdots \leq X_{(N)}$ to denote the ordered sample values, that is, the *order statistics* of the sample, of which $X_{(k+1)}$ is the sample median if $N = 2k + 1$. If $N = 2k$, the sample median is given by $(X_{(k)} + X_{(k+1)})/2$.

Example 2.16. Extending (2.9) to higher dimensions implies that if we have a random vector $(X_1, X_2, \ldots, X_N, S)$ with the joint density $f(x_1, x_2, \ldots, x_N, s)$ and we want to design an estimator for S in the form $T(X_1, X_2, \ldots, X_N)$ such that the mean square error

$$E\{(S - T(X_1, X_2, \ldots, X_N)^2\} \tag{2.12}$$

is minimized, then $T(X_1, X_2, \ldots, X_N)$ is given by the conditional expectation

$$T(X_1, X_2, \ldots, X_N) = E\{S | X_1, X_2, \ldots, X_N\}.$$

In general, the conditional expectation may be very difficult to compute.

We can simplify the problem by seeking a minimum mean square estimator that has a specific form. The simplest case is to require that T is a linear function of $X_1, X_2, \ldots, X_N$. Thus we need to determine $a_1, a_2, \ldots, a_N$ such that

$$E\{(S - T(X_1, X_2, \ldots, X_N))^2\} = E\{(S - (a_1 X_1 + a_2 X_2 + \cdots + a_N X_N))^2\}$$

is minimized. Now, by using the fact that the expectation is a linear operation, we obtain

$$E\{(S - (a_1 X_1 + a_2 X_2 + \cdots + a_N X_N))^2\} = E\{S^2\} - 2\sum_{i=1}^{N} a_i E\{SX_i\} + \sum_{i=1}^{N}\sum_{j=1}^{N} a_i a_j E\{X_i X_j\}.$$

This is a quadratic form in the variables a_i and, e.g., by differentiating with respect to a_i one can show (Exercise) that the minimizing values satisfy

$$
\begin{aligned}
r_{11}a_1 &+ r_{12}a_2 &+ \cdots &+ r_{1N}a_N &= r_1 \\
r_{21}a_1 &+ r_{22}a_2 &+ \cdots &+ r_{2N}a_N &= r_2 \\
&&\vdots&&\\
r_{N1}a_1 &+ r_{N2}a_2 &+ \cdots &+ r_{NN}a_N &= r_N,
\end{aligned}
\tag{2.13}
$$

where $r_{ij} = E\{X_i X_j\}$ form the correlation matrix of the observations and $r_i = E\{SX_i\}$ form the correlation vector between S and the observations.

There can be additional constraints placed on the coefficients a_i. For example, it is often required that the estimator should be *location invariant*, i.e., $X'_n = X_n + c$ implies $E\{Y'\} = E\{Y\} + c$, which leads to the constraint $\sum_{i=1}^{N} a_i = 1$ (Exercise).

Example 2.17. Assume that the relation between S and X_i, $i = 1, 2, \ldots, N$ is of the following form:

$$X_i = S + N_i, \quad i = 1, 2, \ldots, N, \tag{2.14}$$

where the N_is are mutually independent zero mean random variables with variances equal to σ_i^2, $i = 1, 2, \ldots, N$. Assume also that S is independent of the N_is and has zero mean and variance σ_S^2. The relation (2.14) can be interpreted so that we have "noisy" observations of the random variable S. The assumptions made on N_i mean that the "noise" is "white"; it is uncorrelated with the signal and also different noise components are uncorrelated with each other. Straightforward computations show that the correlation matrix can be written as

$$\mathbf{R} = \sigma_S^2 \mathbf{J} + \lceil \sigma_1^2, \sigma_2^2, \ldots, \sigma_N^2 \rfloor,$$

where $\lceil \sigma_1^2, \sigma_2^2, \ldots, \sigma_N^2 \rfloor$ denotes the diagonal matrix with diagonal entries $\sigma_1^2, \sigma_2^2, \ldots, \sigma_N^2$ and $\mathbf{J}$ denotes the matrix with all elements equal to one. Writing $\mathbf{1} = (1, 1, \ldots, 1)^T$ and $\mathbf{a} = (a_1, a_2, \ldots, a_N)^T$ equation (2.13) can be expressed as

$$(\sigma_S^2 \mathbf{J} + \lceil \sigma_1^2, \sigma_2^2, \ldots, \sigma_N^2 \rfloor)\mathbf{a} = \sigma_S^2 \mathbf{1},$$

from which the minimum mean square error solution becomes

$$\mathbf{a} = \left(\mathbf{J} + \left\lceil \frac{\sigma_1^2}{\sigma_S^2}, \frac{\sigma_2^2}{\sigma_S^2}, \ldots, \frac{\sigma_N^2}{\sigma_S^2} \right\rfloor \right)^{-1} \mathbf{1}.$$

It is easy to see (Exercise) that if all the σ_i^2s are equal, all samples will get the same weight, and in the general case more weight is given to those X_is having a small variance in their noise term N_i.

Example 2.18. The mean square error is a natural error criterion and it is especially suitable when the observations obey a Gaussian distribution. However, it is not good in the case of impulsive noise where large deviations occur often. This is because in this case the minimization of the quadratic term makes the estimator too heavily influenced by the worst observations; the ones corrupted by impulses. The effect of outliers can be alleviated by using a less rapidly increasing function. A natural choice is the *mean absolute error*. Thus, we seek the estimator $T(X_1, X_2, \ldots, X_N)$ for S minimizing

$$E\{|S - T(X_1, X_2, \ldots, X_N)|\},$$

where the expectation is again taken with respect to the joint density $f(x_1, x_2, \ldots, x_N, s)$. Now,

$$
\begin{aligned}
&E\{|S - T(X_1, X_2, \ldots, X_N)|\} \\
&= \int_{-\infty}^{\infty} \cdots \int_{-\infty}^{\infty} \int_{-\infty}^{\infty} |s - T(x_1, x_2, \ldots, x_N)| f(x_1, x_2, \ldots, x_N, s) dx_1 dx_2 \cdots dx_N ds \\
&= \int_{-\infty}^{\infty} \cdots \int_{-\infty}^{\infty} \left\{ \int_{-\infty}^{\infty} |s - T(x_1, x_2, \ldots, x_N)| f(x_1, x_2, \ldots, x_N, s) ds \right\} dx_1 dx_2 \cdots dx_N \\
&= \int_{-\infty}^{\infty} \cdots \int_{-\infty}^{\infty} \left\{ \int_{-\infty}^{\infty} |s - T(x_1, x_2, \ldots, x_N)| f(s|x_1, x_2, \ldots, x_N) ds \right\} \cdot \\
&\qquad\qquad\qquad\qquad \cdot f(x_1, x_2, \ldots, x_N) dx_1 dx_2 \cdots dx_N,
\end{aligned}
$$

where $f(s|x_1, x_2, \ldots, x_N)$ denotes the conditional density of S assuming $(X_1, X_2, \ldots, X_N)$ $= (x_1, x_2, \ldots, x_N)$ and $f(x_1, x_2, \ldots, x_N)$ is the joint density of $X_1, X_2, \ldots, X_N$. It is straightforward to show (Exercise) that for each $(x_1, x_2, \ldots, x_N)$ the inner integral

$$\int_{-\infty}^{\infty} |s - T(x_1, x_2, \ldots, x_N)| f(s|x_1, x_2, \ldots, x_N) ds$$

is minimized if $T(x_1, x_2, \ldots, x_N)$ is chosen to be the median of the conditional random variable S assuming $(X_1, X_2, \ldots, X_N) = (x_1, x_2, \ldots, x_N)$. Thus, the general minimum mean absolute error estimator is the *conditional median*, i.e., the median of the conditional distribution,

$$\mathrm{MED}\{S|X_1, X_2, \ldots, X_N\}. \tag{2.15}$$

Again, it may be very difficult to determine the functional form of (2.15).

Remark 2.8. The conditional median (2.15) needs not have much to do with the sample median. In fact, if the random vector $(X_1, X_2, \ldots, X_N, S)$ has a Gaussian distribution it can happen that the conditional median is the sample mean (Exercise), i.e.,

$$\mathrm{MED}\{S|X_1, X_2, \ldots, X_N\} = \frac{1}{N} \sum_{i=1}^{N} X_i.$$

In the following sections, we investigate some general methods of obtaining proper estimators.

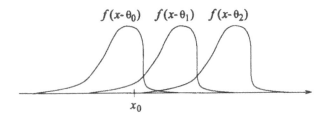

Figure 2.4. Densities $f(\theta_0), f(\theta_1), f(\theta_2)$, and observation x_0.

2.3.2 Maximum Likelihood Estimators

Maximum likelihood estimation is a frequently used method of estimation. To see the basic idea behind maximum likelihood estimation, let us first consider the trivial case of a random sample of size 1.

Example 2.19. Assume that x_0 is a realization of a random variable with density function $f(x - \theta)$ where θ is the unknown location parameter. In Figure 2.4 density functions are depicted for a few values of the parameter θ.

If we were to decide which of the values $\theta_0, \theta_1, \theta_2$ would be most likely, given that we have observed the value x_0, we would say θ_0. This is the principle of maximum likelihood estimation, that is, we choose the value of θ that puts the maximum of $f(x - \theta)$ on the observed value x_0. Thus, the maximum likelihood estimation is essentially an optimization problem. For a random sample with N observations the idea is exactly the same. Only the fact that we work with the high dimensional joint distribution function of the sample values makes it less easy to visualize.

Definition 2.13. Let $\mathbf{x} = (X_1, X_2, \ldots, X_N)$ be a random vector with density function $f(x_1, x_2, \ldots, x_N; \boldsymbol{\theta})$, $\boldsymbol{\theta} \in \Theta$. The function

$$L(\boldsymbol{\theta}: X_1, X_2, \ldots, X_N) = f(X_1, X_2, \ldots, X_N; \boldsymbol{\theta}),$$

considered as a function of $\boldsymbol{\theta}$, is called the *likelihood function*.

Remark 2.9. If $X_1, X_2, \ldots, X_N$ are i.i.d. with density function $f(x; \theta)$, the likelihood function is

$$L(\boldsymbol{\theta}: X_1, X_2, \ldots, X_N) = \prod_{i=1}^{N} f(X_i; \boldsymbol{\theta}).$$

Definition 2.14. The principle of maximum likelihood (point) estimation consists of choosing as an estimate of $\boldsymbol{\theta}$ a $\widehat{\boldsymbol{\theta}}(\mathbf{x})$ that maximizes $L(\boldsymbol{\theta}: X_1, X_2, \ldots, X_N)$, i.e., a mapping $\widehat{\boldsymbol{\theta}} : \mathbf{R}^N \to \mathbf{R}$ satisfying

$$L(\widehat{\boldsymbol{\theta}}: X_1, X_2, \ldots, X_N) = \sup_{\boldsymbol{\theta} \in \Theta} L(\boldsymbol{\theta}; X_1, X_2, \ldots, X_N). \tag{2.16}$$

If a $\widehat{\boldsymbol{\theta}}(\mathbf{x})$ satisfying (2.16) exists, we call it a *maximum likelihood estimator*.

Frequently it is more convenient to work with the logarithm of the likelihood function. This is equivalent, as the logarithm is a strictly increasing function.

Example 2.20.   Consider now *location estimation*, where the parameter appears in the following form

$$f(x;\theta) = f(x - \theta).$$

In other words, the shape and the scale of the underlying density are known; only its location is not known. Assume that $X_1, X_2, \ldots, X_N$ are i.i.d. with normal density

$$f(x;\theta) = \frac{1}{\sqrt{2\pi}\sigma} \exp\{-\frac{(x - \theta)^2}{2\sigma^2}\}.$$

Now, the logarithm of the likelihood function is

$$L(\theta) = -\sum_{i=1}^{N} \frac{(X_i - \theta)^2}{2\sigma^2} - N \ln \sqrt{2\pi}\sigma.$$

Maximizing $L(\theta)$ is equivalent to minimizing

$$\sum_{i=1}^{N}(X_i - \theta)^2,$$

which is obtained by using the argument value (Exercise)

$$\hat{\theta} = \frac{1}{N} \sum_{i=1}^{N} X_i,$$

the mean of the values X_i.

Assume next that the density function of the population is slightly different

$$f(x;\theta) = \frac{\alpha}{2} \exp(-\alpha|x - \theta|),$$

i.e., a Laplace distribution. Now, the logarithm of the likelihood function is given by

$$L(\theta) = -\alpha \sum_{i=1}^{N} |X_i - \theta| + N \ln \alpha - N \ln 2.$$

The maximum likelihood estimator is the operator finding the argument maximizing the likelihood function, i.e., in this case the operator finding θ minimizing

$$\sum_{i=1}^{N} |X_i - \theta|. \tag{2.17}$$

Now, (2.17) is piecewise linear and continuous. It is differentiable except at X_i and the derivative is negative if θ is less than the sample median and positive if θ is greater than the sample median. This means that the sample median minimizes the sum (2.17) and thus, it is the maximum likelihood estimator of location for the Laplace distribution. The detailed proof is left as an exercise. See also Figure 2.5.

Remark 2.10.   These results have a practical meaning in filtering of signals corrupted by additive Gaussian (Laplacian) noise in constant signal regions, where

$$X_i = S + N_i, \quad S \text{ constant}, \quad E(N_i) = 0.$$

In this case, the observed signal values are distributed according to $F(\xi - S)$, where $F(\cdot)$ is the distribution function of the normal (Laplace) distribution. The mean (median) filter can be used as an estimator of the location S, and it is in fact the maximum likelihood estimator of S.

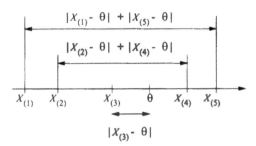

Figure 2.5. The sample median is the value θ minimizing the function (2.17). In this example $N = 5$. The sum $|X_{(1)} - \theta| + |X_{(5)} - \theta|$ finds its minimum and remains constant for $X_{(1)} \leq \theta \leq X_{(5)}$. Similarly, the sum $|X_{(2)} - \theta| + |X_{(4)} - \theta|$ finds its minimum and remains constant for $X_{(2)} \leq \theta \leq X_{(4)}$. Thus, to minimize (2.17), θ has to be $X_{(3)}$, i.e., the sample median.

We have good reasons to believe that the maximum likelihood estimator behaves well for its distribution. However, it may not be a good estimator for some other distributions. Unfortunately, the real data seldom, if ever, follow any simply parametrizable distribution. The maximum likelihood estimator is thus designed only for some idealized situations. As the real data deviate (at least a little) from the model distribution, we need to know whether or not the estimator works reasonably well in the real data. From Tukey [103],

> A tacit hope in ignoring deviations from ideal models was that they would not matter; the statistical procedures which were optimal under the strict model would still be approximately optimal under the approximate model. Unfortunately, it turned out that this hope was often drastically wrong; even mild deviations often have much larger effects than were anticipated by most statisticians.

This is referred to as the *robustness of an estimator.* An estimator is called robust if it performs reasonably well for a wide variety of possible distributions even though it may not be the optimal one for a particular distribution [15]. This is exactly what we need in filtering applications. As the window size is relatively small, the data inside it deviate from the model distribution quite a lot and in various ways when we go through the whole signal. Our filter should be robust to operate successfully throughout the signal. In the following sections we consider some approaches towards robustness. Other approaches can be found, e.g., in [30, pp. 113–116] and in the references therein.

2.3.3 M-Estimators

We saw above that computing the maximum likelihood estimator for a parameter was achieved by finding the argument corresponding to the minimum of the negative of the joint distribution at the observed point $(X_1, X_2, \ldots, X_N)$. In particular the maximum likelihood estimator $\hat{\theta}$ for location minimized

$$\sum_{i=1}^{N}(X_i - \theta)^2$$

for Gaussian data and

$$\sum_{i=1}^{N}|X_i - \theta|$$

for Laplacian data.

M-estimators generalize this in the sense that an M-estimate is found by minimizing a suitable function [39].

Definition 2.15.   Let ρ be a function defined on $\mathbf{R} \times \Theta$. The estimator defined by

$$\hat{\theta} = \arg \min_{\theta \in \Theta} \sum_{i=1}^{N} \rho(X_i, \theta)$$

is called an M-estimator.

Thus the choice $\rho(x, \theta) = (x - \theta)^2$ leads to the sample mean and the choice $\rho(x, \theta) = |x - \theta|$ leads to the sample median. The useful property of M-estimators is that the robustness can be controlled by tailoring the function ρ while maintaining the general nature of the maximum likelihood estimators. The name "M-estimator" comes from "generalized *M*aximum likelihood."

If ρ possesses the partial derivative $\psi(x, \theta) = \frac{\partial}{\partial \theta}\rho(x, \theta)$, then the estimate satisfies the equation

$$\sum_{i=1}^{N} \psi(x_i, \theta) = 0.$$

When we are estimating the location parameter, we usually write $\rho(x, \theta) = \rho(x - \theta)$ and $\psi(x, \theta) = \psi(x - \theta)$. Thus, then the ρ and ψ are functions of only one argument. The most important robust M-estimator that is obtained by carefully designing the ρ-function is the Huber estimator defined by

$$\rho_{\text{Huber}}(x) = \begin{cases} x^2/2, & \text{for } |x| \leq b, \\ b|x| - b^2/2, & \text{for } |x| > b, \end{cases}$$

from which it follows that

$$\psi_{\text{Huber}}(x) = \begin{cases} x, & \text{for } |x| \leq b, \\ b\,\text{sign}(x), & \text{for } |x| > b. \end{cases}$$

2.3.4   *L*-Estimators

Another class of estimators that has found a widespread use is the class of *L-estimators*.

Definition 2.16.   *L*-estimators are of the form

$$\hat{\theta} = \sum_{i=1}^{N} a_i X_{(i)},$$

where a_i are some fixed coefficients.

The name "*L*-estimators" comes from "*L*inear combinations of order statistics". By varying the coefficients (which are usually normalized: $\sum_{i=1}^{N} a_i = 1$) one obtains many useful estimators.

Example 2.21.

(a) Taking $a_i = 1/N$, $i = 1, 2, \ldots, N$, we get

$$\hat{\theta} = \frac{1}{N} \sum_{1=1}^{N} X_i,$$

the *sample mean.*

(b) Taking

$$a_i = \begin{cases} 1, & \text{if } i = k + 1, \\ 0, & \text{otherwise,} \end{cases}$$

where $N = 2k + 1$, we get

$$\hat{\theta} = \text{MED}\{X_1, X_2, \ldots, X_N\},$$

the *sample median.*

(c) Taking $\alpha = j/N$, where $0 \leq j \leq N/2$ is an integer and

$$a_i = \begin{cases} \frac{1}{(1-2\alpha)N}, & \text{if } \alpha N \leq i \leq N - \alpha N, \\ 0, & \text{otherwise,} \end{cases}$$

we obtain

$$\hat{\theta} = \frac{1}{(1 - 2\alpha)N} \sum_{i=\alpha N}^{N - \alpha N} X_i,$$

the *α-trimmed mean* of $X_1, X_2, \ldots, X_N$. This intuitively appealing estimator where we discard αN smallest and αN largest observations has in fact been used for centuries.

(d) Taking $1 \leq j \leq N/2$ and

$$a_i = \begin{cases} 1/2, & \text{if } i = j \text{ or } i = N + 1 - j, \\ 0, & \text{otherwise,} \end{cases}$$

we get

$$\hat{\theta} = \frac{1}{2}(X_{(j)} + X_{(N+1-j)}),$$

the *quasi-midrange*, and if $j = 1$, then

$$\hat{\theta} = \frac{1}{2}(X_{(1)} + X_{(N)}),$$

that is, the *midrange.* At first this estimator seems to discard most of the information, but if, for instance, we only know that the sample values fall uniformly on an interval (a, b), then the midrange is the optimal estimator in the maximum likelihood sense for the center point of the interval $(a + b)/2$ (Exercise).

2.3.5   *R*-Estimators

R-estimators were originally derived from rank tests [37]. The name "*R*-estimator" comes from "*R*ank of an observation X_i". The rank of an observation X_i is denoted by $R(X_i)$ and is given by

$$X_i = X_{(R(X_i))}, \quad i = 1, 2, \ldots, N,$$

i.e., the rank of X_i in the ordered sequence.

R-estimators are usually derived in an implicit way that is useful for the analysis of their general properties. Here we use an alternative definition [42, 66].

Definition 2.17. Assign the weights

$$w_{jk} = \frac{d_{N-k+j}}{\sum_{i=1}^{N} i d_i}$$

to each of the $n(n+1)/2$ averages $(X_{(j)}+X_{(k)})/2, j \leq k$. The R-estimator is the median of the discrete distribution that assigns the probability w_{jk} to each average $(X_{(j)} + X_{(k)})/2$.

Example 2.22.
  (a) Take $d_1 = d_2 = \cdots = d_{N-1} = 0$, $d_N = 1$. Then $\sum i d_i = N$ and

$$w_{jk} = \begin{cases} 1/N, & \text{if } j = k, \\ 0, & \text{otherwise.} \end{cases}$$

Thus the distribution assigns the weight $1/N$ to each X_i, $i = 1, 2, \ldots, N$ meaning that this choice of the weights gives the sample median.
  (b) Take $d_1 = d_2 = \cdots = d_N = 1$. Then the distribution assigns the weight $2/(N(N+1))$ to each of the averages $(X_{(j)} + X_{(k)})/2$ and so

$$\hat{\theta} = \text{MED}\{(X_{(j)} + X_{(k)})/2 : 1 \leq j \leq k \leq N\}.$$

This location estimator is called the *Hodges-Lehmann estimator*.

2.3.6   Scale Estimation

Sometimes it is not enough to estimate only the location of a distribution, but we need also to estimate the scale or spread of the distribution. For example, an estimate of the scale may be needed for the construction of a good location estimator.

The model for the scale is simply

$$F(x; \theta) = F(x/\theta), \tag{2.18}$$

where the parameter space is $\Theta = (0, \infty)$.

The maximum likelihood estimator for scale is, in principle, straightforward to compute from (2.18) but requires exact knowledge of the distribution like in the location estimation. For the Gaussian distribution we obtain (Exercise)

$$\hat{\theta} = \left(\frac{1}{N} \sum_{i=1}^{N} X_i^2\right)^{1/2},$$

which is also called the *biased sample standard deviation*. It is obvious from its form that it is quite sensitive to outliers.

The ψ-function of an M-estimate for scale is of the form

$$\psi(x; \theta) = \psi(x/\theta),$$

and a good choice of ψ will lead to a useful robust estimator for scale. The choice

$$\psi(x) = \psi_{\text{MAD}}(x) = \text{sign}(|x| - G^{-1}(3/4))$$

where $G(x)$ is the distribution function of normalized Gaussian distribution (with zero mean and unity variance) leads to the *(standardized) median of absolute deviations from the median (MAD)*:

$$\hat{\theta} = 1.483 \text{MED}\{|X_i - \text{MED}\{X_i : 1 \leq i \leq N\}| : 1 \leq i \leq N\}.$$

The scaling parameter 1.483 makes $\hat{\theta}$ to converge to the true value for the normal distribution.

Often we use the MAD without the scaling parameter, and when we have no additional knowledge of the distribution, this is perhaps the most useful indicator of the scale. It is interesting to note that the first person to have used the MAD was probably Gauss [27].

2.4 Some Useful Distributions

In this section we briefly describe the properties of distributions that are useful in the analysis of nonlinear filters. These results will be used in Chapter 4 where we investigate the statistical properties of nonlinear filters. Some of the concepts that are introduced in the following are fairly difficult. The derivations below are quite terse and the reader may wish to consult a book on probability theory and statistics, e.g., [116].

In general, the *gamma function* $\Gamma(z)$ is defined for complex numbers z whose real part is positive by the integral

$$\Gamma(z) = \int_0^\infty x^{z-1}e^{-x}dx.$$

We will only use $\Gamma(z)$ for real positive z. It is easy to see that

$$\Gamma(z) = (z-1)\Gamma(z-1), \quad \text{for} \ \ z > 1,$$

from which it follows that for a positive integer

$$\Gamma(z) = (z-1)!.$$

If $z > 0$ but not an integer, that is, $z = i + u$, where $i \in \mathbf{Z}_+$ and $0 < u < 1$, we have

$$\Gamma(z) = (z-1)(z-2)\cdots u\Gamma(u).$$

In particular $\Gamma(\frac{1}{2}) = \sqrt{\pi}$.

The *gamma distribution* is related to the sums of squares of Gaussian random variables and to waiting times.

Definition 2.18. Let $\mu > 0$. A random variable whose probability density function is

$$f_G(x) = \frac{x^{\mu-1}e^{-x}}{\Gamma(\mu)}, \quad x > 0$$

is said to have the gamma distribution with parameter μ.

Remark 2.11. The rth moment of a gamma distributed random variable is

$$E\{X^r\} = \int_0^\infty x^r \frac{x^{\mu-1}e^{-x}}{\Gamma(\mu)}dx = \frac{1}{\Gamma(\mu)}\int_0^\infty x^{r+\mu-1}e^{-x}dx = \frac{\Gamma(\mu+r)}{\Gamma(\mu)},$$

from which it follows that $E\{X\} = \text{Var}\{X\} = \mu$.

Another distribution which is closely connected to some of the nonlinear filters to be studied later is the *beta distribution*.

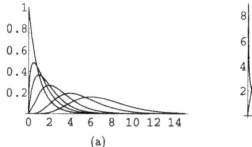
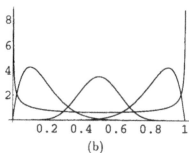

Figure 2.6. Density functions of random variables having (a) gamma distribution, $\mu = 1, 1.5, 2, 3, 5, 7$ in the order of the peaks from left to right; (b) beta distribution, $(\alpha, \beta) = (0.5, 0.5)$ corresponds to the U-shaped density, $(\alpha, \beta) = (10, 10)$ corresponds to the density with peak at 0.5, $(\alpha, \beta) = (2, 10)$ corresponds to the density with peak close to 0.1, and $(\alpha, \beta) = (10, 2)$ corresponds to the density with peak close to 0.9.

Definition 2.19. A random variable whose density is given by

$$f_B(x) = \frac{\Gamma(\alpha + \beta)}{\Gamma(\alpha)\Gamma(\beta)} x^{\alpha-1}(1-x)^{\beta-1}, \quad 0 < x < 1, \tag{2.19}$$

where α and β are positive, is said to have the beta distribution with parameters α and β.

The moments of the beta distribution are easily computed from (2.19)

$$E\{X^r\} = \frac{\Gamma(\alpha + \beta)\Gamma(\alpha + r)}{\Gamma(\alpha + \beta + r)\Gamma(\alpha)},$$

giving then

$$E\{X\} = \frac{\alpha}{\alpha + \beta}, \quad \text{Var}\{X\} = \frac{\alpha\beta}{(\alpha + \beta)^2(\alpha + \beta + 1)}.$$

Some densities of gamma and beta distributed random variables are plotted in Figure 2.6.

The k-dimensional analog of the beta distribution is the *k-dimensional Dirichlet distribution*:

Definition 2.20. Let $\alpha_1, \alpha_2, \ldots, \alpha_k, \beta > 0$ and let $(X_1, X_2, \ldots, X_k)$ be a random vector that has the density

$$f(x_1, x_2, \ldots, x_k) = \frac{\Gamma(\alpha_1 + \cdots + \alpha_k + \beta)}{\Gamma(\alpha_1) \cdots \Gamma(\alpha_k)\Gamma(\beta)} x_1^{\alpha_1 - 1} \cdots x_k^{\alpha_k - 1}(1 - x_1 - \cdots - x_k)^{\beta - 1} \tag{2.20}$$

at any point in the set $\{(x_1, x_2, \ldots, x_k) : x_i > 0, \ i = 1, 2, \ldots, k, \sum x_i < 1\}$ and zero outside. A distribution having the density (2.20) is called the $k-$dimensional Dirichlet distribution $D(\alpha_1, \alpha_2, \ldots, \alpha_k; \beta)$.

Note that for $k = 1$, (2.20) is the beta density.

One can verify (Exercise) that the moments, that is, $E\{X_1^{r_1} X_2^{r_2} \cdots X_k^{r_1}\}$, of the $k-$dimensional Dirichlet distribution are

$$E\{X_1^{r_1} X_2^{r_2} \cdots X_k^{r_1}\} = \frac{\Gamma(\alpha_1 + r_1) \cdots \Gamma(\alpha_k + r_k)\Gamma(\alpha_1 + \cdots + \alpha_k + \beta)}{\Gamma(\alpha_1 + \cdots + \alpha_k + \beta + r_1 + \cdots + r_k)\Gamma(\alpha_1 + \cdots + \alpha_k)},$$

from which, e.g., means, variances and covariances of the X_is can be easily computed (Exercise).

The following result for random vectors with the Dirichlet distribution will be used when analyzing so-called order statistic filters in Chapter 4. It shows that summing segments of a Dirichlet distributed random vector gives a Dirichlet distributed random vector.

If $(X_1, X_2, \ldots, X_k)$ is a random vector having the k-dimensional Dirichlet distribution $D(\alpha_1, \ldots, \alpha_k; \beta)$, then the random vector $(Z_1, Z_2, \ldots, Z_s)$, where

$$
\begin{cases}
Z_1 = \sum_{i=1}^{k_1} X_i, \\[2mm]
Z_2 = \sum_{i=k_1+1}^{k_1+k_2} X_i, \\[2mm]
\vdots \\[2mm]
Z_s = \sum_{i=k_1+k_2+\cdots+k_{s-1}+1}^{k_1+k_2+\cdots+k_s} X_i,
\end{cases}
\tag{2.21}
$$

and $k_1 + k_2 + \cdots + k_s \le k$ has the s-dimensional Dirichlet distribution $D(\gamma_1, \ldots, \gamma_s; \delta)$, where

$$
\begin{cases}
\gamma_1 = \alpha_1 + \alpha_2 + \cdots + \alpha_{k_1}, \\[2mm]
\vdots \\[2mm]
\gamma_s = \alpha_{k_1+\cdots+k_{s-1}+1} + \cdots + \alpha_{k_1+\cdots+k_s}, \\[2mm]
\delta = \alpha_{k_1+\cdots+k_s+1} + \cdots + \alpha_k + \beta.
\end{cases}
\tag{2.22}
$$

Example 2.23. Consider a random vector (X_1, X_2, X_3) with the Dirichlet distribution $D(2, 2, 3; 2)$. Its density is according to (2.20) given by

$$
f(x_1, x_2, x_3) = \frac{\Gamma(9)}{\Gamma(2)\Gamma(2)\Gamma(3)\Gamma(2)} x_1 x_2 x_3^2 (1 - x_1 - x_2 - x_3),
$$

for $x_i > 0$, $x_1 + x_2 + x_3 < 1$, and zero otherwise. Now take $Z_1 = X_1 + X_2$, $Z_2 = X_3$. From (2.21) it follows that the random vector (Z_1, Z_2) has the Dirichlet distribution $D(4, 3; 2)$:

$$
f(z_1, z_2) = \frac{\Gamma(9)}{\Gamma(4)\Gamma(3)\Gamma(2)} z_1^3 z_2^2 (1 - z_1 - z_2)^2 = \frac{8!}{3!2!1!} z_1^3 z_2^2 (1 - z_1 - z_2)^2.
$$

The connection of Dirichlet distributions to order statistics can be seen by applying the following transformation to the Dirichlet distributed random vector $(X_1, X_2, \ldots, X_k)$

$$
\begin{cases}
Y_1 = X_1, \\[2mm]
Y_2 = X_1 + X_2, \\[2mm]
\vdots \\[2mm]
Y_k = X_1 + X_2 + \cdots + X_k.
\end{cases}
$$

Since the Jacobian of the transformation is unity, we have the following density for $(Y_1, \ldots Y_k)$:

$$
f(y_1, \ldots, y_k) = \frac{\Gamma(\alpha_1 + \cdots + \alpha_k + \beta)}{\Gamma(\alpha_1)\cdots\Gamma(\alpha_k)\Gamma(\beta)} y_1^{\alpha_1-1} (y_2 - y_1)^{\alpha_2-1} \cdots (y_k - y_{k-1})^{\alpha_k-1} (1 - y_k)^{\beta-1}
$$

for $0 < y_1 < y_2 < \cdots < y_k < 1$. It is convenient to call this density the *ordered k-dimensional Dirichlet density* $D^*(\alpha_1, \ldots, \alpha_k; \beta)$. It can be shown that if $(Y_1, \ldots Y_k)$ is a random vector having the ordered k-dimensional Dirichlet distribution $D^*(\alpha_1, \ldots, \alpha_k; \beta)$, then the marginal distribution of $(Y_{k_1}, Y_{k_1+k_2}, \ldots, Y_{k_1+k_2+\cdots+k_s})$ is the s-dimensional ordered Dirichlet distribution $D^*(\gamma_1, \gamma_2, \ldots, \gamma_s; \delta)$ where the parameters are as in (2.22). This property is useful in deriving the distributions of order statistics.

3

1001 SOLUTIONS

IF THOU WILT DEIGN THIS FAVOUR, FOR THY MEED;
A THOUSAND HONEY SECRETS SHALT THOU KNOW
William Shakespeare, "Venus and Adonis", 1593

THE MOST EXCITING PHRASE TO HEAR IN SCIENCE,
THE ONE THAT HERALDS THE MOST DISCOVERIES,
IS NOT "EUREKA!" (I FOUND IT!) BUT "THAT'S FUNNY..."
Isaac Asimov, 1920–1992

The last three decades have witnessed a tremendous growth in the literature dealing with nonlinear filters designed to solve the filtering problem discussed in Chapter 1. Without exaggeration one can claim that hundreds of filter classes have been developed and analyzed to some extent. The literature on nonlinear filters is nowadays enormous, and an attempt to do justice to everything would prove futile. Thus, we restrict discussions to fundamental core material without attempting to cover everything or to be fair to all contributors in the field. In this section, we review what we consider to be some of the main classes of these filters. We have chosen these filters to illustrate the wide range of alternatives for solving the aforementioned problems. After reading this chapter the reader should understand the depth of the problems and the fact that they cannot be solved all at the same time. In other words, if one of the problems has been satisfactorily solved, some of them remain unsolved at the same time, or even worse—new more serious problems are born. It is typical for many of the filters studied so far that they have been designed for a particular application where they perform very well but they normally perform rather poorly outside their own field.

The aim of this section is to familiarize the reader with each filter in a couple of ways:

- By describing a way of thinking which could have led to the invention of the filter. In other words, a possible motivation behind the filter is discussed. The reader can easily localize the idea in the text by finding the sign ⬚ in the margin.

- By giving an exact definition of the filter indicated by the sign ⬚.

- By giving a straightforward algorithm for realizing the filter. This enables the reader to program the filter easily and thus use his own test signals to verify its behavior.

- By studying its impulse and step response. The impulse response is studied using as a filter input the sequence $\{\ldots, 0, 0, 0, a, 0, 0, 0, \ldots\}$, where a is either positive or negative. An ideal step edge is represented by the sequence $\{\ldots, a, a, a, b, b, b, \ldots\}$, where $a \neq b$. The reader is encouraged to consider the two-dimensional case on his own.

- By showing a representative set of example signals illustrating how the filter really works. The same set of example signals is used throughout this chapter allowing direct comparisons between different filter classes. We strongly encourage the reader to carefully compare different filters by using these signals, as we believe that the human eye is the best tool for measuring signal fidelity. The filter parameters have not been optimized in any case and the same window size is used throughout this section. If some suggestions were found in the literature, then those parameters were chosen accordingly. In other cases they were chosen in the best way according to our experience with the filters under consideration. We comment on the results very briefly and mainly leave it to the readers to perform comparisons between the filters.

- By providing a set of supporting exercises. These exercises have been selected to contain the essentials omitted from the basic text.

- By providing the basic references to the filters. This book is not aimed to be a handbook in the sense of providing a thorough list of references, though.

From now on we assume that the size, N, of the moving window is odd, i.e., $N = 2k+1$ ($k \neq 1$). The set of observation samples inside the moving window will be denoted by $\{X_1, X_2, \ldots, X_N\}$. The sample around which the window is centered and the filtering operation performed is denoted by X^*. The algorithms are mainly given for the two-dimensional case. However, it is straightforward to modify these algorithms to other dimensions as well, provided that the filters can be defined to cover other dimensions. Sometimes to emphasize that the window is two-dimensional we call the window a $z \times z$ window, for which $z \times z = N$. In the one-dimensional filtering examples the window length is 11. The filter is thus using as an input, in addition to the center sample X^*, five samples before and after it. In the two-dimensional filtering examples a 5×5 window is used, i.e., $z = 5$, $k = 12$, and $N = 25$.

In this chapter we have altogether 22 sections, each containing several filter types. As the reader will see, some of the filter classes are clearly related to each other, whereas between certain filter classes, there does not seem to be any unifying underlying idea outside the uttermost one–to filter out the noise. Thus, taxonomies grouping different filters in some natural classes would be desirable. However, none of the suggested taxonomies has gained an uncontested status. It might even be impossible to produce such a taxonomy. Now, we illustrate two possible classifications without any preference for either:

Taxonomy I	Taxonomy II
Generalizations of Linear Filters	*Estimator Interpretation*
3.1 Trimmed Mean Filters	3.1 Trimmed Mean Filters
3.2 Other Trimmed Mean Filters	3.2 Other Trimmed Mean Filters
3.14 Nonlinear Mean Filters	3.3 L-Filters
3.19 Polynomial Filters	3.4 C-Filters
Filters Based on Ordering	3.5 Weighted Median Filters
3.3 L-Filters	3.6 Ranked-Order and
3.4 C-Filters	Weighted Order Statistic Filters
3.5 Weighted Median Filters	3.11 M-Filters
3.6 Ranked-Order and	3.12 R-Filters
Weighted Order Statistic Filters	3.14 Nonlinear Mean Filters
3.7 Multistage Median Filters	3.15 Stack Filters
3.8 Median Hybrid Filters	3.16 Generalizations of Stack Filters
3.10 Rank Selection Filters	*Geometric Interpretation*
3.15 Stack Filters	3.17 Morphological Filters
3.16 Generalizations of Stack Filters	3.18 Soft Morphological Filters
Morphological Filters	*Others*
3.17 Morphological Filters	3.7 Multistage Median Filters
3.18 Soft Morphological Filters	3.8 Median Hybrid Filters
Others	3.9 Edge-Enhancing Selective Filters
3.9 Edge-Enhancing Selective Filters	3.10 Rank Selection Filters
3.11 M-Filters	3.13 Weighted Majority with
3.12 R-Filters	Minimum Range Filters
3.13 Weighted Majority with	3.19 Polynomial Filters
Minimum Range Filters	3.20 Data-Dependent Filters
3.20 Data-Dependent Filters	3.21 Decision-Based Filters
3.21 Decision-Based Filters	3.22 Iterative, Cascaded,
3.22 Iterative, Cascaded,	and Recursive Filters
and Recursive Filters	

3.1   Trimmed Mean Filters

3.1.1   Principles and Properties

In Chapter 1 we recognized that both the median filter and the mean filter had desirable as well as undesirable properties. Most notably, the median filter discarded impulses well, but the case of additive Gaussian noise was more problematic for it. In turn, the performance of the mean filter was superior to that of the median in removing additive Gaussian noise but deteriorated dramatically with impulsive type noise. Therefore, good compromises between the median and the mean might lead to filters with good behavior in situations where both Gaussian and impulsive noise are present. Perhaps the simplest way to obtain this kind of compromise is to use *trimmed means*. Trimmed means probably date back to the prehistory of statistics since the idea of trimming out some suspicious looking samples is very obvious. Possibly the first article about trimmed means was published in 1821 [3]. The author is not known; he might have been Gergonne. The idea behind a trimmed mean is to reject the most probable outliers—some of the very smallest and very largest values, and after rejection to average the rest. Here we refer to Mendeleev (1895) [76]; (see [34]):

> I use... [the following] method to evaluate ...: I divide all the numbers into three, if possible equal, groups (if the number of observations is not divisible by three, the greatest number is left in the middle group): those of greatest magnitude, those of medium magnitude, and those of smallest magnitude: the mean of the middle group is considered the most probable ...

This idea is playing an important role in certain sport performances, like figure skating, where several judges evaluate the individual (style, technique, theatrical, ...) performances and some judges might be biased. To diminish the effect of the biased evaluations, the greatest and the smallest marks are rejected.

Trimmed mean filters have been considered, e.g., in [28, 14, 64]. The first form of a trimmed mean filter, the (r, s)-*fold trimmed mean filter*, is obtained by sorting the samples and then omitting altogether $r+s$ samples, $X_{(1)}, X_{(2)}, \ldots, X_{(r)}$ and $X_{(N-s+1)}, X_{(N-s+2)}, \ldots, X_{(N)}$. Then the output is generated by averaging the remaining samples, i.e.,

$$\text{TrMean}(X_1, X_2, \ldots, X_N; r, s) = \frac{1}{N - r - s} \sum_{i=r+1}^{N-s} X_{(i)}.$$

An immediate modification of the (r, s)-trimmed mean filter is the (r, s)-*fold Winsorized mean filter*, where the values of the r smallest samples are replaced by $X_{(r+1)}$ and the values of the s largest samples are replaced by $X_{(N-s)}$, yielding

$$\text{WinMean}(X_1, X_2, \ldots, X_N; r, s) = (1/N)(rX_{(r+1)} + \sum_{i=r+1}^{N-s} X_{(i)} + sX_{(N-s)}).$$

If the numbers of the trimmed elements from both ends are equal, $r = s$, the amount of the trimmed elements is often specified by the proportion of the elements trimmed. This proportion is typically denoted by $\alpha = j/N$, where $0 \leq j \leq N/2$ is an integer. The number of αN samples are trimmed at each end. The corresponding filters are called the α-*trimmed mean filter* and the α-*Winsorized mean filter*.

(r, s)-fold trimmed mean filter

Inputs: *NumberOfRows* × *NumberOfColumns* image
　　　　Moving window $W, |W| = N = 2k + 1$
　　　　Natural numbers r, s
Output: *NumberOfRows* × *NumberOfColumns* image

```
for i = 1 to NumberOfRows
    for j = 1 to NumberOfColumns
        place the window W at (i, j)
        store the image values inside W into x = (X₁, X₂, ..., Xₙ)
        sort x, store the result in y = (X₍₁₎, X₍₂₎, ..., X₍ₙ₎)
        let Sum = 0
        for m = r + 1 to N - s
            let Sum = Sum + X₍ₘ₎
        end
        let Output(i, j) = Sum/(N - r - s)
    end
end
```

Algorithm 3.1. Algorithm for the (r, s)-fold trimmed mean filter.

$$\text{TrMean}(X_1, X_2, \ldots, X_N; \alpha) = \frac{1}{N - 2\alpha N} \sum_{i=\alpha N+1}^{N-\alpha N} X_{(i)};$$

$$\text{WinMean}(X_1, X_2, \ldots, X_N; \alpha) = (1/N)(\alpha N \cdot X_{(\alpha N+1)} + \sum_{i=\alpha N+1}^{N-\alpha N} X_{(i)} + \alpha N \cdot X_{(N-\alpha N)}).$$

An algorithm for the (r, s)-fold trimmed mean filter is given in Algorithm 3.1. The other algorithms are easily obtained by minor modifications to the innermost for-loop, the formula of the output, and to inputs.

Clearly, the $(0, 0)$-fold trimmed, $(0, 0)$-fold Winsorized, 0-trimmed, and 0-Winsorized mean filters are the same as the mean filter of the same window size; similarly the (k, k)-fold trimmed, (k, k)-fold Winsorized, 0.5-trimmed, and 0.5-Winsorized mean filters are the same as the median filter of the same window size. When only a few of the samples are trimmed out, the behavior of the trimmed mean filters is close to that of the mean filter. The more trimming is used, the more the filter behavior resembles that of the median filter.

Example 3.1. Let the 5-point input be $x = (2, 7, 9, 20, 5)$. The order of these samples is $X_{(1)} = 2$, $X_{(2)} = 5$, $X_{(3)} = 7$, $X_{(4)} = 9$, and $X_{(5)} = 20$. The output of the

- $(2, 1)$-fold trimmed mean filter is $(7 + 9)/2 = 8$;

- $(1, 2)$-Winsorized mean filter is $(2 \cdot 5 + 3 \cdot 7)/5 = 6.2$;

- 0.2-trimmed mean is $(5 + 7 + 9)/3 = 7$;

- 0.2-Winsorized mean is $(2 \cdot 5 + 7 + 2 \cdot 9)/5 = 7$.

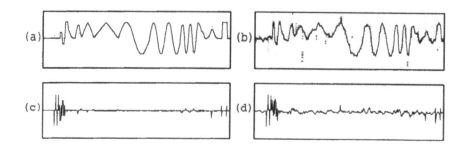

Figure 3.1. One-dimensional signals filtered by the 4/11-trimmed mean filter of window length 11.

Figure 3.2. "Geometrical" filtered by the 10/25-trimmed mean filter of window size 5 × 5.

3.1.2   Impulse and Step Response

An impulse appears at either end of the ordered sequence of the values inside the window. From this it directly follows: the impulse responses of the (Winsorized) trimmed means are zero-valued sequences provided that $r, s, \alpha N \neq 0$. We have already noticed that the median filter preserves step edges completely. In fact, it is the only example of the (Winsorized) trimmed mean filters having this property. All the other trimmed mean filters either transform a step edge into a ramp edge and/or change its position (Exercise). Winsorized means cause also slight jumps before and after the ramps. A rule of thumb can be given: The more trimming, the better the step edge is preserved.

3.1.3   Filtering Examples

From the examples it is easy to see that the α-trimmed mean filter is the expected compromise between the mean and the median filters. The α-trimmed mean filters used in the figures failed to completely preserve the edges but they didn't blur them to the extent that the mean filter did. In the one-dimensional signals we can also see the effect of the rounding of the result which is needed to give the output in the 8-bit form similar to the input.

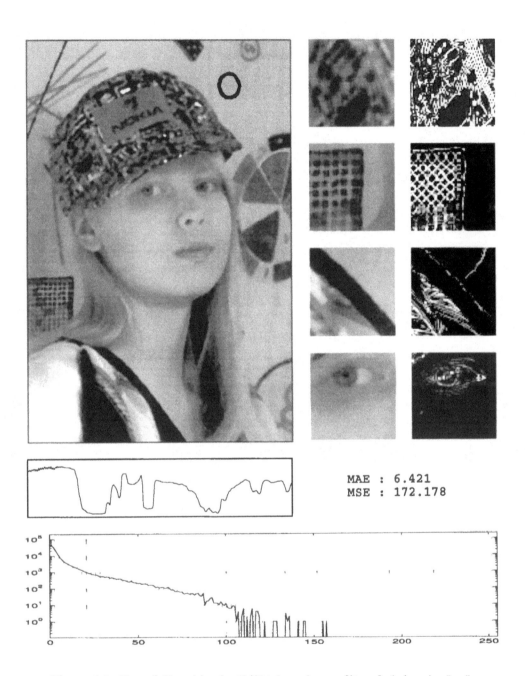

MAE : 6.421
MSE : 172.178

Figure 3.3. "Leena" filtered by the 10/25-trimmed mean filter of window size 5 × 5.

Modified trimmed mean filter

Inputs: *NumberOfRows* × *NumberOfColumns* image
        Moving window $W, |W| = N = 2k + 1$
        Positive real number q
Output: *NumberOfRows* × *NumberOfColumns* image

```
for i = 1 to NumberOfRows
    for j = 1 to NumberOfColumns
        place the window W at (i, j)
        store the image values inside W in x = (X₁, X₂, ..., Xₙ)
        find the median X₍ₖ₊₁₎ of x
        let Sum1 = 0,  Sum2 = 0
        for m = 1 to N
            if |Xₘ − X₍ₖ₊₁₎| ≤ q then
                let Sum1 = Sum1 + Xₘ
                let Sum2 = Sum2 + 1
        end
        let Output(i, j) = Sum1/Sum2
    end
end
```

Algorithm 3.2. Algorithm for the modified trimmed mean filter.

3.2 Other Trimmed Mean Filters

3.2.1 Principles and Properties

In the α-trimmed mean filter, the set of $k - \alpha N$ samples closest to the median $X_{(k+1)}$ is selected from each of the two sets of k samples on either side of the median. The output is given by the average of these samples and the median. In other words, the output is found to be the average of the samples in a window inside the range $[X_{(k+1)} - q_1, X_{(k+1)} + q_2]$, where q_1 and q_2 depend on the data. The *modified trimmed mean filter* [63, 64] is based on the idea to change q_1 and q_2 to be independent of the data. In this filter, a real-valued constant q is first fixed and then all the samples in the range $[X_{(k+1)} - q, X_{(k+1)} + q]$ are averaged, and the average gives the output; see Algorithm 3.2. Therefore, the idea of the modified trimmed means is to exclude those samples in the window being sufficiently far away from the median value. The parameter q is used to classify the samples into two sets: those which are too far from the median and those which are close enough to the median. Note that the number of the samples used in the averaging is now not fixed *a priori* but is dependent on the data.

More formally:

$$\text{MTM}(X_1, X_2, \ldots, X_N; q) = \frac{\sum_{i=1}^{N} a_i X_i}{\sum_{i=1}^{N} a_i},$$

where

$$a_i = \begin{cases} 1, & \text{if } |X_i - X_{(k+1)}| \leq q, \\ 0, & \text{otherwise.} \end{cases} \tag{3.1}$$

Double window modified trimmed mean filter

Inputs: *NumberOfRows* × *NumberOfColumns* image
        Moving window W_1, $|W_1| = N = 2k + 1$
        Moving window W_2, $|W_2| = M, M > N$
        Positive real number q
Output: *NumberOfRows* × *NumberOfColumns* image

```
for i = 1 to NumberOfRows
    for j = 1 to NumberOfColumns
        place the windows W₁ and W₂ at (i,j)
        store the image values inside W₁ in x = (X₁, X₂, ..., Xₙ)
        find the median X₍ₖ₊₁₎ of x
        let Sum1 = 0, Sum2 = 0
        store the image values inside W₂ in y = (X₁, X₂, ..., Xₘ)
        for m = 1 to M
            if |Xₘ − X₍ₖ₊₁₎| ≤ q then
                let Sum1 = Sum1 + Xₘ
                let Sum2 = Sum2 + 1
        end
        let Output(i, j) = Sum1/Sum2
    end
end
```

Algorithm 3.3. Algorithm for the double window modified trimmed mean filter.

Example 3.2. Let the 5-point input be $\mathbf{x} = (2, 7, 9, 20, 5)$, where the median is 7. Now, the output of the modified trimmed mean filter is

- the median 7, for $0 \leq q < 2$;

- $(7 + 9 + 5)/3 = 7$, for $2 \leq q < 5$;

- $(2 + 7 + 9 + 5)/4 = 5.75$, for $5 \leq q < 13$;

- the mean $(2 + 7 + 9 + 20 + 5)/5 = 8.6$, otherwise.

A natural extension of the modified trimmed mean filter is to first compute the median, denoted by *med*, from a relatively small window (size N) and use, as the output, the mean of the samples lying within the interval $[med - q, med + q]$ among the samples in a larger window (size M), centered at the same point. See Algorithm 3.3. The idea behind this filter, *double window modified trimmed mean filter*, [64] is to be able to use large windows to suppress additive Gaussian noise and, at the same time, preserve details by rejecting pixels that are far away from the median of the smaller window.

Example 3.3. Let the 5-point input be $\mathbf{x} = (2, 7, 9, 20, 5)$, and let the three samples in the center belong to the smaller window ($N = 3$) and let $M = 5$. The median of the smaller window is 9. The output of the double window modified trimmed mean filter is

- the median of the smaller window 9, for $0 \leq q < 2$;

K-nearest neighbor filter

Inputs: *NumberOfRows* × *NumberOfColumns* image
        Moving window $W, |W| = N = 2k+1$
        Natural number $K, 1 \leq K \leq N$
Output: *NumberOfRows* × *NumberOfColumns* image

```
for i = 1 to NumberOfRows
    for j = 1 to NumberOfColumns
        place the window W at (i, j)
        store the image values inside W and their differences from
            the value of X* (pixel in (i, j)) in
```
$$x = ((X_1, |X_1 - X^*|), (X_2, |X_2 - X^*|), \ldots, (X_N, |X_N - X^*|))$$
```
        sort x with respect to |X_i - X*|, store the result in
```
$$y = ((X_{(1)}, |X_{(1)} - X^*|), (X_{(2)}, |X_{(2)} - X^*|), \ldots, (X_{(N)}, |X_{(N)} - X^*|)$$
```
        let Sum = 0
        for m = 1 to K
            let Sum = Sum + X_(m)
        end
        let Output(i, j) = Sum/K
    end
end
```

Algorithm 3.4. Algorithm for the K-nearest neighbor filter.

- $(7 + 9)/2 = 8$, for $2 \leq q < 4$;

- $(7 + 9 + 5)/3 = 7$, for $4 \leq q < 7$;

- $(2 + 7 + 9 + 5)/4 = 5.75$, for $7 \leq q < 11$;

- the mean $(2 + 7 + 9 + 20 + 5)/5 = 8.6$, otherwise.

Clearly, by choosing q to be very small, the (double window) modified trimmed mean filters behave like median filters, whereas for very large q values they behave like mean filters. Thus, the parameter q provides us a tool to control the behavior of the filter. Selection of q is typically done utilizing available *a priori* information about the heights of the edges and the variance of the noise distribution. This is discussed in more detail in Subsection 3.2.2.

The *K-nearest neighbor filter* [24] can also be understood as a trimmed mean filter. In this filter, the output is given by the mean of the $K, 1 \leq K \leq N$, samples whose values are closest to the value of the central sample X^* inside the filter window; see Algorithm 3.4. Note that X^* is always one of these samples as its distance from itself is zero, the smallest possible. In some cases it may happen that there are several different samples whose distances from X^* are equal and by taking them all into account more than K values would be chosen. In this case, some tie-breaking rule must be applied. The underlying idea in this method is to preserve edges by using only those neighbors in the averaging that most likely belong to the same region as X^*. However, it is obvious that this method is unable to remove impulsive noise.

Example 3.4. Let the 5-point input be $\mathbf{x} = (2, 7, 9, 20, 5)$. The center sample $X^* = 9$.

- For $K = 1$ the sample whose value is closest to 9 is the only one taken into account. This sample is clearly the sample X^* itself. Thus, the output is now 9. In fact, for $K = 1$ the K-nearest neighbor filter is always the identity filter.

- For $K = 2$ the two samples whose values are closest to 9, i.e., 9 and 7, are averaged, giving the output 8.

- For $K = 3$ the output is the average of $9, 7, 5$, i.e., 7.

- For $K = 4$ the average of $9, 7, 5, 2$, i.e., 5.75.

- For $K = 5$ the output is 8.6, the mean of all the samples.

By changing the rule for a_i in (3.1) into the form

$$a_i = \left\{ \begin{array}{ll} 1, & \text{if } |X_i - X^*| \leq q, \\ 0, & \text{otherwise,} \end{array} \right. \tag{3.2}$$

the *modified nearest neighbor filter* is obtained [64]. If the threshold q is chosen to be twice the standard deviation σ of the noise, this filter is known as the *sigma filter* [61]. Thus, the sigma filter is a special case of the modified nearest neighbor filter. By the definition, in the modified nearest neighbor filtering, those samples with values close enough to the value of the central sample X^* are averaged. Again, X^* is always one of these samples. The parameter q is used to trim out the values which most likely come from a different population than the central pixel. Clearly, the modified nearest neighbor filter suffers from the same problem as the K-nearest neighbor filter: its ability to remove impulsive noise is practically nonexistent.

Example 3.5. Let the 5-point input be $\mathbf{x} = (2, 7, 9, 20, 5)$. Now, the center sample is $X^* = 9$. The output of the modified nearest neighbor filter is

- the center sample 9, for $0 \leq q < 2$;

- $(7 + 9)/2 = 8$, for $2 \leq q < 4$;

- $(7 + 9 + 5)/3 = 7$, for $4 \leq q < 7$;

- $(7 + 9 + 5 + 2)/4 = 5.75$, for $7 \leq q < 11$;

- 8.6, the mean of all the samples, otherwise.

3.2.2 Impulse and Step Response

In the (double window) modified trimmed mean filtering the median is first calculated and then the values which are far away from the median are rejected. An impulse is clearly such a value, thus, the impulse response of the modified trimmed mean filter is a zero-valued sequence. The nearest neighbor filters, in turn, cannot completely remove an impulse as the center sample X^* is always in the averaging process even when X^* is an impulse. If $K < N$ or q is smaller than the height of the impulse, then the impulse is discarded from the averaging when X^* is not an impulse. Thus, in these cases the impulse will not be spread.

Modified nearest neighbor filter

Inputs: *NumberOfRows* × *NumberOfColumns* image
        Moving window $W, |W| = N = 2k + 1$
        Positive real number q
Output: *NumberOfRows* × *NumberOfColumns* image

```
for i = 1 to NumberOfRows
    for j = 1 to NumberOfColumns
        place the window W at (i, j)
        store the image values inside W in x = (X₁, X₂, ..., Xₙ)
        let Sum1 = 0, Sum2 = 0
        for m = 1 to N
            if |Xₘ − X*| ≤ q then
                let Sum1 = Sum1 + Xₘ
                let Sum2 = Sum2 + 1
        end
        let Output(i, j) = Sum1/Sum2
    end
end
```

Algorithm 3.5. Algorithm for the modified nearest neighbor filter.

To preserve step edges, the parameter q of the modified filters must satisfy $q < H$, where H is the height of the edge. This follows from the fact that the samples on the wrong side of the edge (compared to $X_{(k+1)}$) have distance H from the median, and by choosing $q < H$ they are rejected, yielding the perfect step preservation. If $q \geq H$, then the step edge will be transformed into a ramp edge. The K-nearest neighbor filters preserve step edges if $K \leq k + 1$. This is easily deduced from the fact that in the window there are always at least k samples having the same value than the central sample. The modified nearest neighbor filters preserve step edges if $q < H$. Then all the samples not having the same value as the central sample are discarded from the averaging.

3.2.3   Filtering Examples

Clearly, the modified trimmed mean filters inherit properties similar to those of the median filter—the output is always close to the median value. However, by taking the mean value of the points close enough to the median provides an improved smoothing of Gaussian noise compared to the median filter.

The double window modified trimmed mean filter is quite successful in many ways in the examples. One reason for this is that the size and the shape of the smaller window suits well certain parts of signal. From Figure 3.5 (b) and (d) we can see that as the smaller window (7 in this case) is more sensitive to the impulses than the larger window (11 in Figure 3.4) more impulses can survive the double window modified trimmed mean filtering compared with modified trimmed mean filtering of the same window size.

We do not show the results of filtering by the modified nearest neighbor filters because they are similar to the results of the K-nearest neighbor filtering. As is natural to expect, the K-nearest neighbor filter cannot remove impulses, but it is able to smooth Gaussian noise somehow. Furthermore, it preserves details well, especially for small K.

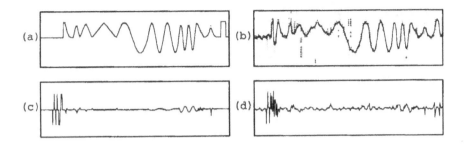

Figure 3.4. One-dimensional signals filtered by the modified trimmed mean filter of window length 11, $q = 24$.

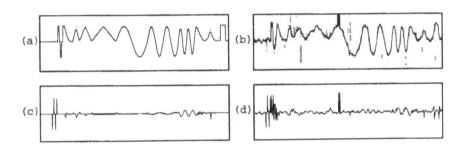

Figure 3.5. One-dimensional signals filtered by the double window modified trimmed mean filter with smaller window length 7, larger window length 11, and $q = 24$.

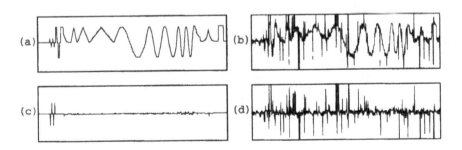

Figure 3.6. One-dimensional signals filtered by the K-nearest neighbor filter of window length 11, $K = 4$.

Figure 3.7. "Geometrical" filtered by the modified trimmed mean filter of window size 5×5, $q = 35$.

Figure 3.8. "Geometrical" filtered by the double window modified trimmed mean filter with smaller window size 3×3, larger window size 5×5, and $q = 30$.

Figure 3.9. "Geometrical" filtered by the K-nearest neighbor filter of window size 5×5, $K = 9$.

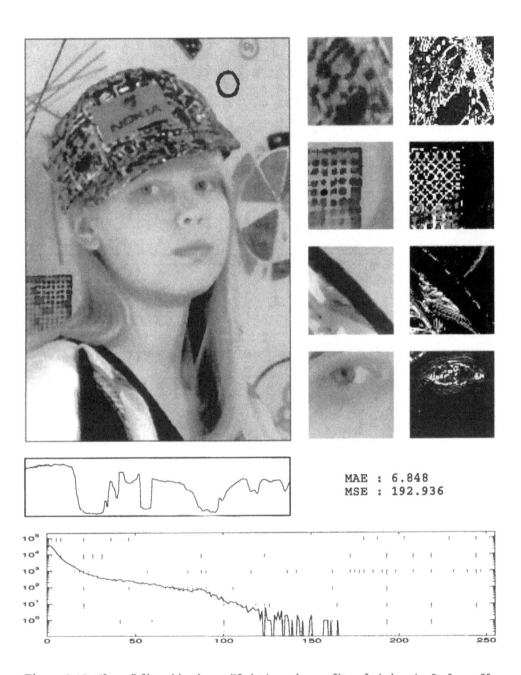

MAE : 6.848
MSE : 192.936

Figure 3.10. "Leena" filtered by the modified trimmed mean filter of window size 5×5, $q = 35$.

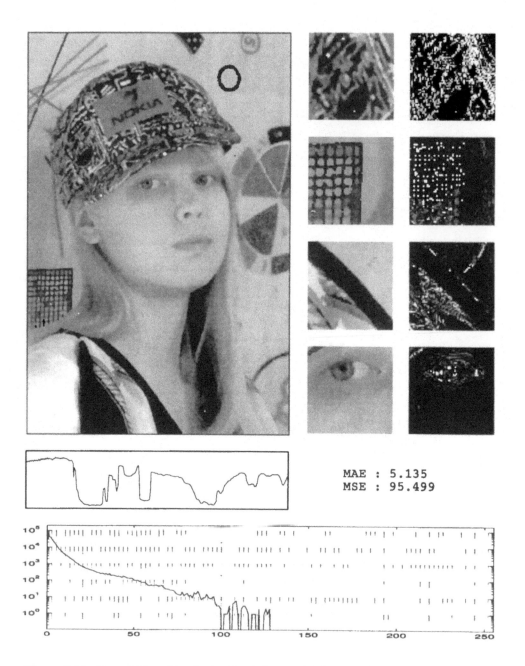

MAE : 5.135
MSE : 95.499

Figure 3.11. "Leena" filtered by the double window modified trimmed mean filter with smaller window size 3 × 3, larger window size 5 × 5, and $q = 30$.

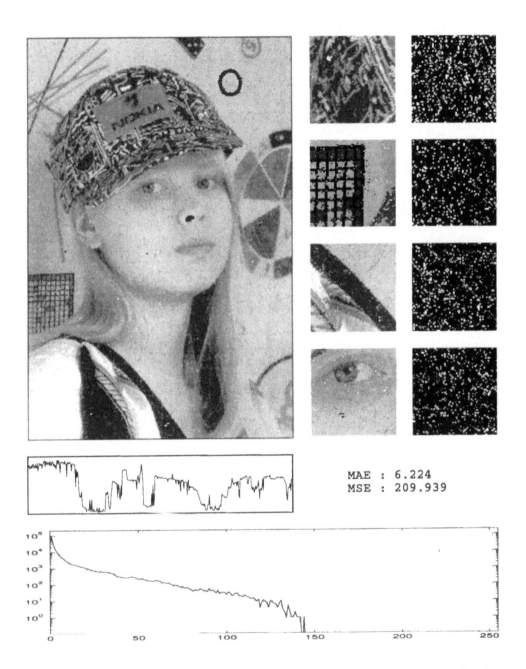

MAE : 6.224
MSE : 209.939

Figure 3.12. "Leena" filtered by the K-nearest neighbor filter of window size 5×5, $K = 9$.

3.3 L-Filters

3.3.1 Principles and Properties

Among the problems of the median filter that we have encountered are streaking and edge jitter. An important group of nonlinear filters attempting to ameliorate these drawbacks is the group of *L-filters*, earlier also called *order statistic filters* [17, 18, 28, 63, 64]. Linear combinations of order statistics have long been recognized by statisticians as having robust and often optimal properties for estimating population parameters for i.i.d. observations. For instance, Lloyd [70] showed that the location parameter can be more efficiently estimated with a linear combination of ordered samples than by the classical sample mean. The corresponding mean square error will always be smaller than or equal to that obtained with the sample mean or the sample median.

L-filters are running estimators making a compromise between a pure nonlinear operation (ordering) and a pure linear operation (weighting) (cf. Figure 3.13). Each output point is obtained as a weighted sum of ordered data values in the moving window; see Algorithm 3.6. The set of weights determines the characteristics of the filters. Let us denote the weight vector of an L-filter by $\mathbf{a} = (a_1, a_2, \ldots, a_N)$ and the input vector by $\mathbf{x} = (X_1, X_2, \ldots, X_N)$. Then the output of the L-filter is given by

$$L(X_1, X_2, \ldots, X_N; \mathbf{a}) = \sum_{i=1}^{N} a_i X_{(i)}.$$

Example 3.6. Let the 5-point input be $\mathbf{x} = (2, 7, 9, 20, 5)$. The ordered input vector is $(2, 5, 7, 9, 20)$. We obtain the following weight vector and L-filter output pairs:

$\mathbf{a} = (0.5, 0, 0, 0, 0.5),$ \qquad $L(2, 7, 9, 20, 5; \mathbf{a}) = 0.5 \cdot 2 + 0.5 \cdot 20 = 11;$
$\mathbf{a} = (0, 0.1, 0.8, 0.1, 0),$ \qquad $L(2, 7, 9, 20, 5; \mathbf{a}) = 0.1 \cdot 5 + 0.8 \cdot 7 + 0.1 \cdot 9 = 7;$
$\mathbf{a} = (-0.1, 0, 1.2, 0, -0.1),$ \quad $L(2, 7, 9, 20, 5; \mathbf{a}) = -0.1 \cdot 2 + 1.2 \cdot 7 - 0.1 \cdot 20 = 6.2.$

The first coefficients give the *midpoint*. The second coefficients define an L-filter with window length 3, as two of the weights equal to zero.

The median, mean, (r, s)-fold-trimmed means, α-trimmed means, and Winsorized means are all also L-filters (Exercise). A natural way to generate L-filters is obtained by using [30]

$$a_i = \frac{\int_{\frac{i-1}{N}}^{\frac{i}{N}} h(\lambda) d\lambda}{\int_0^1 h(\lambda) d\lambda}, \tag{3.3}$$

where $h : [0, 1] \rightarrow \mathbf{R}$ satisfies $\int_0^1 h(\lambda) d\lambda \neq 0$. (Exercise: Why normalization?)

L-filters satisfying

$$\sum_{i=1}^{N} a_i = 1 \tag{3.4}$$

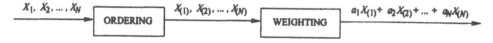

Figure 3.13. L-filtering operation.

L-filter

Inputs: *NumberOfRows* × *NumberOfColumns* image
Moving window $W, |W| = N = 2k + 1$
Weight vector $a = (a_1, a_2, \ldots, a_N)$
Output: *NumberOfRows* × *NumberOfColumns* image

```
for i = 1 to NumberOfRows
    for j = 1 to NumberOfColumns
        place the window W at (i, j)
        store the image values inside W in x = (X₁, X₂, ..., Xₙ)
        sort x, store the result in y = (X₍₁₎, X₍₂₎, ..., X₍ₙ₎)
        let Sum = 0
        for m = 1 to N
            let Sum = Sum + aₘ · X₍ₘ₎
        end
        let Output(i, j) = Sum
    end
end
```

Algorithm 3.6. Algorithm for the *L*-filter.

are sometimes called *smoothing L-filters*. This term refers to the property that for a constant-valued signal, the output has also the same constant value if and only if the weights satisfy (3.4).

By using other ordering rules which generate novel permutations of the input samples prior to weighting we obtain some generalized order statistic filters [71] (see also Exercises).

The great advantage of the *L*-filter class is that for a known noise distribution it is possible to choose the filter weights in such a way that it becomes the optimal filter in the mean square error sense. This will be described in detail in Chapter 4.

3.3.2 Impulse and Step Response

The impulse response of an *L*-filter is trivially a zero-valued sequence given $a_1 = a_N = 0$. Consider now the edge preservation properties of *L*-filters. It is easy to see that the discontinuity between two flat portions in the input signal in the step edge case will, in general, be smeared by *L*-filters. The extent to which the edges are smeared naturally depends on the weights. In order to obtain good edge preservation properties, the data from the other side of the edge should be treated as impulsive noise components and be discarded from the output calculation. When the window is located exactly at the edge, k samples should be understood as impulses. The standard median filter is the only *L*-filter having this property and it is the only *L*-filter which can preserve an ideal edge.

3.3.3   Filtering Examples

The weights for the one-dimensional signals are

$$\begin{array}{cccccc}
a_1, a_{11} & a_2, a_{10} & a_3, a_9 & a_4, a_8 & a_5, a_7 & a_6 \\
-0.01886 & 0.01809 & 0.02907 & 0.10965 & 0.20795 & 0.30820
\end{array} \tag{3.5}$$

and for the two-dimensional signals they are

$$\begin{array}{ccccccc}
a_1, a_{25} & a_2, a_{24} & a_3, a_{23} & a_4, a_{22} & a_5, a_{21} & a_6, a_{20} & a_7, a_{19} \\
0.00550 & 0.00335 & -0.00427 & -0.00101 & -0.00008 & 0.00065 & 0.00314
\end{array}$$
$$\begin{array}{cccccc}
a_8, a_{18} & a_9, a_{17} & a_{10}, a_{16} & a_{11}, a_{15} & a_{12}, a_{14} & a_{13} \\
0.01064 & 0.02907 & 0.06499 & 0.11835 & 0.17195 & 0.19541
\end{array} \tag{3.6}$$

The weights for the two-dimensional signals are the optimal ones in the mean square error sense for the Laplace distribution. Also the weights for the one-dimensional case suit well in relatively long-tailed distributed noise cases. These weights are not surprising as most of the weight is located in the central order statistics, so that the resulting filter is close to the median filter and to the α-trimmed mean filters with α close to $1/2$. Visual comparisons using the corresponding figures justify this fact.

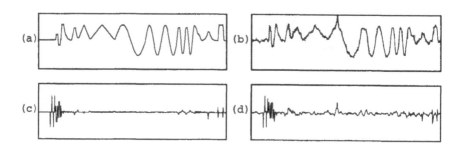

Figure 3.14. One-dimensional signals filtered by the L-filter of window length 11 and weights shown in (3.5).

Figure 3.15. "Geometrical" filtered by the L-filter of window size 5×5 and weights shown in (3.6).

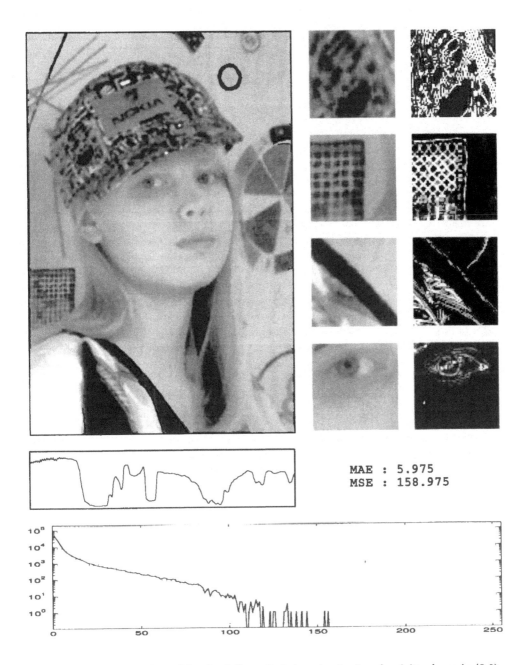

MAE : 5.975
MSE : 158.975

Figure 3.16. "Leena" filtered by the L-filter of window size 5×5 and weights shown in (3.6).

3.4 C-Filters (Ll-Filters)

3.4.1   Principles and Properties

L-filters (as well as median filters) discard the temporal-order (spatial-order) information of the data completely. This causes problems, e.g., in detail preservation, and the problems become more serious with the increase of the window size. The following example clearly shows how order statistic based filters suffer from the limitation that they do not take into account the temporal information of the signals. Consider the following samples:

$$\mathbf{x}_1 = (10, 17, 20, 17, 10, 0, -10, -17, -20, -17, -10, 0, 10)$$
$$\mathbf{x}_2 = (-20, -17, -17, -10, -10, 0, 0, 10, 10, 10, 17, 17, 20),$$

which are observed by sampling a sinusoid ($\mathbf{x}_1$) and an increasing ramp ($\mathbf{x}_2$). Although these signals have very distinct and different patterns, their corresponding sorted vectors are identical. In fact the same sorted vector could be produced by any permutation of $\mathbf{x}_1$. The number of permutations of 13 numbers is 13! which is over $6 \cdot 10^9$. Naturally, in this case not all the values of the signals are different, meaning that the actual number of different permutations is less than 13!. As the temporal-order information is discarded, these filters are also unable to recover the spectral characteristics of the signal.

A method of incorporating both temporal-order and rank-order information of the input sequence is the *combination filter* or *C-filter* [45, 86]. This filter is a special case of the Ll-filters [84]. C-filters include both linear FIR (finite impulse response) filters and L-filters and a C-filter can be understood, e.g., in the following ways:

- A C-filter is an L-filter with the coefficient of the $X_{(i)}$ dependent on its temporal position in the window.

- A C-filter is an FIR filter with the coefficient of the X_i dependent on its rank within the window.

The output of the C-filter with $N \times N$ coefficient matrix $\mathbf{C} = [c(i, j)]$ and inputs $(X_1, X_2, \ldots, X_N)$ is given by

$$C(X_1, X_2, \ldots, X_N; \mathbf{C}) = \sum_{i=1}^{N} c(R(X_i), i) X_i,$$

where $R(X_i)$ is the rank of the sample X_i.

Thus, the actual coefficients depend on the ordering of the input samples. Every ordering (permutation) of them picks the corresponding coefficients from $\mathbf{C}$ and then the weighted sum of the inputs gives the output. This is a serious underlying problem with the C-filter: one has to carefully choose coefficients for every permutation in order to avoid creating some undesired artifacts into the filtering results. These artifacts may be due, e.g., to the lack of gain (sum of the weights is less than 1). One possible approach to solving this problem is to normalize the output at every window position by the sum of those N weights used in the output computation. Still, the matrix must be designed with care (see Exercises). Thus, the normalized C-filter is defined by (see Algorithm 3.7)

$$\text{NormC}(X_1, X_2, \ldots, X_N; \mathbf{C}) = \frac{\sum_{i=1}^{N} c(R(X_i), i) X_i}{\sum_{i=1}^{N} c(R(X_i), i)}.$$

C-filter

Inputs: *NumberOfRows* × *NumberOfColumns* image
Moving window $W, |W| = N = 2k + 1$
Weight matrix $C = [c(i,j)]_{N \times N}$
Output: *NumberOfRows* × *NumberOfColumns* image

```
for i = 1 to NumberOfRows
    for j = 1 to NumberOfColumns
        place the window W at (i, j)
        store the image values inside W in x = (X₁, X₂, ..., Xₙ)
        sort x, store the result in y = (X₍₁₎, X₍₂₎, ..., X₍ₙ₎)
        for every element Xᵢ of x find its position R(Xᵢ) in y,
            store the result in r = (R(X₁), R(X₂), ..., R(Xₙ))
        let Sum1 = 0, Sum2 = 0
        for m = 1 to N
            Sum1 = Sum1 + c(R(Xₘ), m) · Xₘ
            Sum2 = Sum2 + c(R(Xₘ), m)
        end
        let Output(i, j) = Sum1/Sum2
    end
end
```

Algorithm 3.7. Algorithm for the normalized *C*-filter.

Naturally, we must be careful with the ties, since then two or more samples have the same value but they should get different ranks. This can be solved, e.g., in Algorithm 3.7 by first finding the value X_i (and its position) in the sorted vector $\mathbf{y}$ and then changing the value X_i in $\mathbf{y}$ into some other value that could not be in $\mathbf{y}$, i.e., some value outside the range of possible inputs.

Example 3.7. Let the 5-point input be $\mathbf{x} = (2, 7, 9, 20, 5)$. Then the ranks of the samples are $(1, 3, 4, 5, 2)$ and the output is $2 \cdot c(1,1) + 7 \cdot c(3,2) + 9 \cdot c(4,3) + 20 \cdot c(5,4) + 5 \cdot c(2,5)$. For the coefficient matrices

$$
C_1 = \begin{pmatrix}
\mathbf{0.00} & 0.05 & 0.05 & 0.05 & 0.00 \\
0.10 & 0.15 & 0.15 & 0.15 & \mathbf{0.10} \\
0.50 & \mathbf{0.65} & 0.80 & 0.65 & 0.50 \\
0.10 & 0.15 & \mathbf{0.15} & 0.15 & 0.10 \\
0.00 & 0.05 & 0.05 & \mathbf{0.05} & 0.00
\end{pmatrix}, C_2 = \begin{pmatrix}
\mathbf{-0.10} & -0.10 & 0.00 & -0.10 & -0.10 \\
0.20 & 0.25 & 0.55 & 0.25 & \mathbf{0.20} \\
0.50 & \mathbf{0.95} & 1.80 & 0.95 & 0.50 \\
0.20 & 0.25 & \mathbf{0.55} & 0.25 & 0.20 \\
-0.10 & -0.10 & 0.00 & \mathbf{-0.10} & -0.10
\end{pmatrix}
$$

the outputs are for the unnormalized case $C(2, 7, 9, 20, 5; C_1) = 0.00 \cdot 2 + 0.65 \cdot 7 + 0.15 \cdot 9 + 0.05 \cdot 20 + 0.10 \cdot 5 = 7.4$ and $C(2, 7, 9, 20, 5; C_2) = -0.10 \cdot 2 + 0.95 \cdot 7 + 0.55 \cdot 9 - 0.10 \cdot 20 + 0.20 \cdot 5 = 10.4$. In the normalized case we have to normalize the previous outputs by the sum of the weights, i.e., by $0.00 + 0.65 + 0.15 + 0.05 + 0.10 = 0.95$ for C_1 and by $-0.10 + 0.95 + 0.55 - 0.10 + 0.20 = 1.5$ for C_2. Thus, the outputs for the normalized case are $NormC(2, 7, 9, 20, 5; C_1) = 7.4/0.95 \approx 7.79$ and $NormC(2, 7, 9, 20, 5; C_2) = 10.4/1.5 \approx 6.93$. The boldfaced numbers in the matrices denote the coefficients used in this permutation.

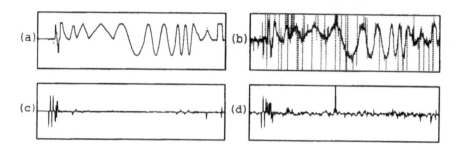

Figure 3.17. One-dimensional signals filtered by a normalized C-filter of window length 11.

3.4.2 Impulse and Step Response

If the first and Nth rows of the matrix $\mathbf{C}$ are zero, then the impulse response of the C-filter is a zero-valued sequence. The impulse response is completely determined by the values in these rows. The C-filter does not preserve the step edge in general as the output procedure includes averaging operations.

3.4.3 Filtering Examples

The results are shown for the normalized C-filters. The weight matrices are generated in the following simple way, where our aim is to incorporate also some temporal-order (spatial-order) information into the filtering. The weight matrix is mainly formed by the same vectors used in the L-filtering examples. Only one column of the matrix differs from the other columns, that of the sample X^{*}. The elements of this column $\mathbf{b}$ are equal in the one-dimensional vector to

b_1, b_{11}	b_2, b_{10}	b_3, b_9	$b_4, \ldots, b_8$
-0.01886	0.01809	0.02907	4.0

and so the weight matrix is given by $\mathbf{C} = (\mathbf{a}^T, \mathbf{a}^T, \mathbf{a}^T, \mathbf{a}^T, \mathbf{a}^T, \mathbf{b}^T, \mathbf{a}^T, \mathbf{a}^T, \mathbf{a}^T, \mathbf{a}^T, \mathbf{a}^T)$, where the coefficients of the vector $\mathbf{a}$ are given in (3.5).

In the two-dimensional examples the coefficients of the vector $\mathbf{a}$ are given in (3.6) and the elements of the column $\mathbf{b}$ corresponding to the center sample X^{*} are

b_1, b_{25}	b_2, b_{24}	b_3, a_{23}	b_4, b_{22}	b_5, b_{21}	b_6, b_{20}	b_7, b_{19}	$b_8, \ldots, b_{18}$
0.00550	0.00335	-0.00427	-0.00101	-0.00008	0.00065	0.00314	$6.0,$

thus, the weight matrix is given by $\mathbf{C} = (\overbrace{\mathbf{a}^T, \ldots, \mathbf{a}^T}^{12\ \text{times}}, \mathbf{b}^T, \overbrace{\mathbf{a}^T, \ldots, \mathbf{a}^T}^{12\ \text{times}})$. Hence, the temporal-order (spatial-order) information included in the filter is of the form of essentially increased weights for the sample X^{*} when it is not one of the extreme order statistics. Thus, in these cases this sample will be dominating in the averaging, resulting in improved detail preservation. In the case of an impulsive X^{*}, the weights corresponding to the resulting ordering are that of the original L-filter indicating that the impulse will be efficiently removed.

From all the examples the improved detail preservation properties can be seen.

Figure 3.18. "Geometrical" filtered by a normalized C-filter of window size 5×5.

3.5 Weighted Median Filters

3.5.1 Principles and Properties

In Chapter 1 the streaking problem of median filters was encountered. Another serious problem discussed was the loss of small details from the images. The main reason for this is that the median filter uses only the rank-order information from the input data, and discards the original temporal-order (spatial-order) information of the data as explained in the previous section. In other words, in the median filter, each sample inside the filter window has the same influence on the filter output. To achieve new properties, one might want to give more emphasis to signal samples in specific window positions, e.g., to the center sample X^*—the sample which we are filtering at the moment. Thus, we want to emphasize the samples that for some reason are supposed to be more reliable, and the emphasis is obtained by weighting them more heavily. This idea has led to the development of the *weighted median filters* [19, 44]. Reference [120] provides a thorough survey of weighted median filtering. The concept itself was already known in the 19th century [16].

The relationship between the median filter and the weighted median filters is comparable to that between the FIR filters and the mean filter. Let a_i be the weight for the sample X_i, $1 \le i \le N$. The output of a weighted median filter can be found by duplicating each input sample a_i times and choosing the median from the resulting array of $\sum a_i$ samples. In a similar way to the FIR filters being much more flexible than the mean filters, weighted medians are much more flexible than the median filters.

Let a_i denote the weight corresponding to the input X_i. Assume first that the weights are positive integers. Furthermore, denote the *duplication operation* or the *repetition operation* by $\lozenge$, that is,

$$r \lozenge x = \overbrace{x, \ldots, x}^{r \text{ times}}.$$

A *multiset* is a collection of objects, where the repetition of objects is allowed. For example, $\{1, 1, 1, 2, 3, 3\} = \{3 \lozenge 1, 2, 2 \lozenge 3\}$ is a multiset.

The output of the weighted median filter with weights $\mathbf{a} = (a_1, a_2, \ldots, a_N)$ and inputs $\mathbf{x} = (X_1, X_2, \ldots, X_N)$ is given by

$$\text{WeightMed}(X_1, X_2, \ldots, X_N; \mathbf{a}) = \text{MED}\{a_1 \lozenge X_1, a_2 \lozenge X_2, \ldots, a_N \lozenge X_N\}. \tag{3.7}$$

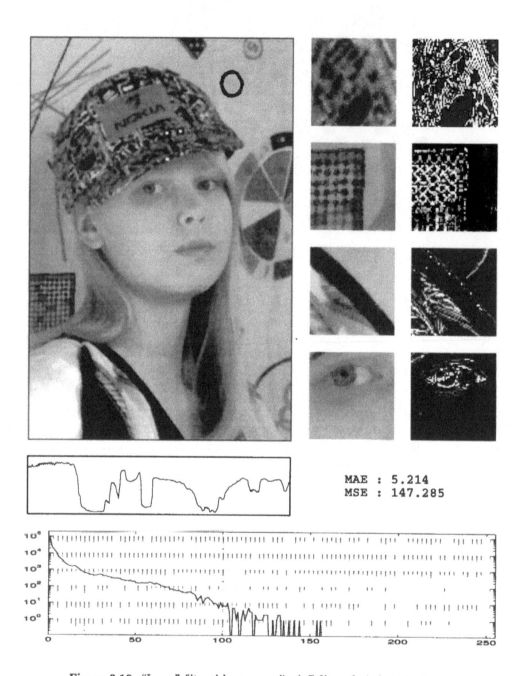

MAE : 5.214
MSE : 147.285

Figure 3.19. "Leena" filtered by a normalized C-filter of window size 5×5.

Weighted median filter

Inputs: *NumberOfRows* × *NumberOfColumns* image
         Moving window $W, |W| = N = 2k + 1$
         Weight vector $\mathbf{a} = (a_1, a_2, \ldots, a_N)$
Output: *NumberOfRows* × *NumberOfColumns* image

let $HalfSum = \sum_{i=1}^{N} a_i/2$
for $i = 1$ to *NumberOfRows*
     for $j = 1$ to *NumberOfColumns*
         place the window W at (i, j)
         store the image values inside W and the corresponding
             weights in $\mathbf{x} = ((X_1, a_1), (X_2, a_2), \ldots, (X_N, a_N))$
         sort $\mathbf{x}$ with respect to X_is, store the result in
             $\mathbf{y} = ((X_{(1)}, a_{(1)}), (X_{(2)}, a_{(2)}), \ldots, (X_{(N)}, a_{(N)}))$
         let $Sum = 0,\ m = 1$
         repeat
             let $Sum = Sum + a_{(m)}$
             let $m = m + 1$
         until $Sum \geq HalfSum$
         let $Output(i, j) = X_{(m-1)}$
     end
end

Algorithm 3.8. Algorithm for the weighted median filter.

Another definition of the weighted median filter allows positive non-integer weights to be used:

$$\text{WeightMed}(X_1, X_2, \ldots, X_N; \mathbf{a}) = \arg\min_{\beta} \sum_{i=1}^{N} a_i |X_i - \beta|, \qquad (3.8)$$

in other words, the weighted median of $(X_1, X_2, \ldots, X_N)$ is the value β minimizing

$$\sum_{i=1}^{N} a_i |X_i - \beta|.$$

The output is guaranteed to be one of the samples X_i since the right-hand side of (3.8) is piecewise linear and convex if $a_i \geq 0$ for all i.

The output of the weighted median filter for positive real weights can be calculated as follows: Sort the samples inside the filter window, add up the corresponding weights from either end of the sorted set until their sum $\geq \frac{1}{2} \sum_{i=1}^{N} a_i$, then the output of the weighted median filter is the sample corresponding to the last weight. This is illustrated in Algorithm 3.8. The same algorithm can also be applied even though some of the weights are negative, even though the intuitive meaning of the resulting operation is not clear. Furthermore, then we encounter some problems (see Exercises). It is important to notice that by using this method independent of the weights we have to only order N samples, not $\sum_{i=1}^{N} a_i$ samples.

Example 3.8. Let the weights of a length 5 weighted median filter and the input be

$$\mathbf{a} = (0.1, 0.2, 0.3, 0.2, 0.1) \quad \text{and} \quad \mathbf{x} = (2, 7, 9, 20, 5).$$

After sorting we obtain the sorted input set with corresponding weights

$$((2, 0.1), (5, 0.1), (7, 0.2), (9, 0.3), (20, 0.2)).$$

Now, we have to add up the first four of the weights $0.1 + 0.1 + 0.2 + 0.3$ to have their sum greater than $0.45 = \sum a_i / 2$. Thus, the filter output is 9.

The weight parameters allow fine-tuning of the performance of the weighted median filter for specific applications. We must emphasize that there are only a finite number of weighted median filters for a given finite window size even though there exist an infinite number of real weight vectors [20, 81]. Furthermore, it has been shown that every weighted median filter with real-valued weights is identical with a weighted median filter with integer-valued weights [81].

3.5.2   Impulse and Step Response

It is trivial to notice that the impulse response of a weighted median filter is a zero-valued sequence if the weights a_i satisfy $a_i < (\sum_{j=1}^{N} a_j)/2$ for every i. A sufficient condition for step preservation is the use of symmetrical weights, that is, $a_i = a_{N-i+1}$ for all i. It is also obvious that no weighted median filter blurs step edges, since the output of a weighted median filter is always one of the samples inside the moving window. For non-symmetrical weights the location of the step edge may be transformed.

3.5.3   Filtering Examples

The weights used for the one-dimensional signals are $\mathbf{a}_1 = (3, 3, 3, 7, 11, 19, 11, 7, 3, 3, 3)$ and $\mathbf{a}_2 = (1, 1, 1, 1, 1, 5, 1, 1, 1, 1, 1)$; in the two-dimensional examples the weights are:

$$\mathbf{a}_1 = \begin{pmatrix} 1 & 1 & 2 & 1 & 1 \\ 1 & 3 & 4 & 3 & 1 \\ 2 & 4 & 11 & 4 & 2 \\ 1 & 3 & 4 & 3 & 1 \\ 1 & 1 & 2 & 1 & 1 \end{pmatrix} \quad \text{and} \quad \mathbf{a}_2 = \begin{pmatrix} 1 & 1 & 1 & 1 & 1 \\ 1 & 1 & 1 & 1 & 1 \\ 1 & 1 & 13 & 1 & 1 \\ 1 & 1 & 1 & 1 & 1 \\ 1 & 1 & 1 & 1 & 1 \end{pmatrix}. \tag{3.9}$$

The first weights have been determined to preserve certain structures in the signals, pulses of length 3 in the one-dimensional case and corners in the two-dimensional case. The second ones are examples of so called *center weighted median filters* [50]. These filters give more weight only to the central value of a window. This leads to improved detail preservation properties at the expense of lower noise suppression, especially in the mean square error sense as now some of the impulses may not be removed by the filter.

The improved detail preservation capability of weighted median filters compared with that of the median filter is clearly demonstrated in these examples.

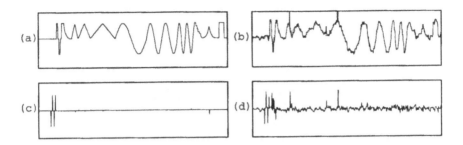

Figure 3.20. One-dimensional signals filtered by the weighted median filter of window length 11 and weights $(3, 3, 3, 7, 11, 19, 11, 7, 3, 3, 3)$.

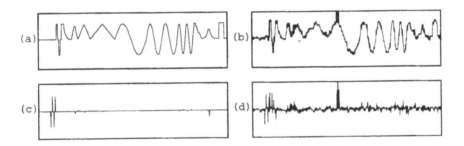

Figure 3.21. One-dimensional signals filtered by the weighted median filter of window length 11 and weights $(1, 1, 1, 1, 1, 5, 1, 1, 1, 1, 1)$.

Figure 3.22. "Geometrical" filtered by the weighted median filter of window size 5×5 and weights a_1 shown in (3.9).

Figure 3.23. "Geometrical" filtered by the weighted median filter of window size 5×5 and weights a_2 shown in (3.9).

3.6   Ranked-Order and Weighted Order Statistic Filters

3.6.1   Principles and Properties

Ranked-order filters are simple modifications of the median filter. The output of the rth ranked-order filter is given by

$$\mathrm{RO}(X_1, X_2, \ldots, X_N; r) = X_{(r)},$$

that is, the rth order statistic of the samples $X_1, X_2, \ldots, X_N$ in the filter window; see Algorithm 3.9. Special cases of this filter class are the median ($r = k + 1$), the maximum operation ($r = N$), and the minimum operation ($r = 1$). The last two cases will be studied in Section 3.17. Ranked-order filters can be used in situations where the noise distribution is not symmetric, e.g., when there are more positive than negative impulses. In this case the use of some low order ranked-order filter ($r < k$) may provide more robust filters compared with, e.g., the median filter. We may say that ranked-order filters adapt better to different noise distributions than the median filters do. The algorithm for ranked-order filters is basically the same as for the median filter; see Algorithm 3.9. The idea of ranked-order filtering is old. The ranked-order filters are based on order statistics, whose literature is voluminous. It is not easy to pick the first reference where some order statistic other than median is used in a filtering application. In the course of the history, these filters have also been known as *percentile filters* and *order filters*.

Obviously, the rth order statistic filters introduce bias towards small values if $r < k + 1$ and towards large values if $r > k + 1$. The strength of the bias depends on the value r and the input distribution. Values of r close to 1 give stronger bias than the values slightly smaller than $k + 1$. Similarly, the closer r is to N, the stronger the bias. The bias tends to be stronger for heavy-tailed distributions. Thus, we must emphasize that the rth order statistic filters should be used with care, since their performance varies for different input distributions. Also, too strong a bias may be visually unpleasing.

Example 3.9.   Let the 5-point input be $\mathbf{x} = (2, 7, 9, 20, 5)$. Then the ordered vector is $(2, 5, 7, 9, 20)$ and the outputs of the rth ranked-order filters are 2 for $r = 1$, 5 for $r = 2$, 7 for $r = 3$, 9 for $r = 4$, and 20 for $r = 5$.

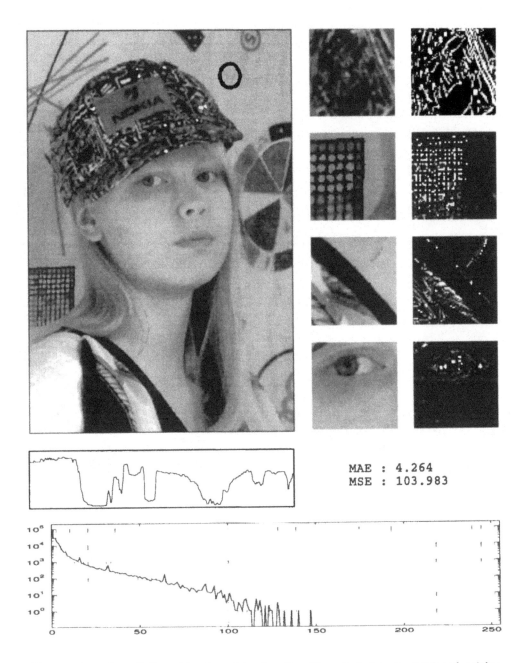

MAE : 4.264
MSE : 103.983

Figure 3.24. "Leena" filtered by the weighted median filter of window size 5 × 5 and weights a_1 shown in (3.9).

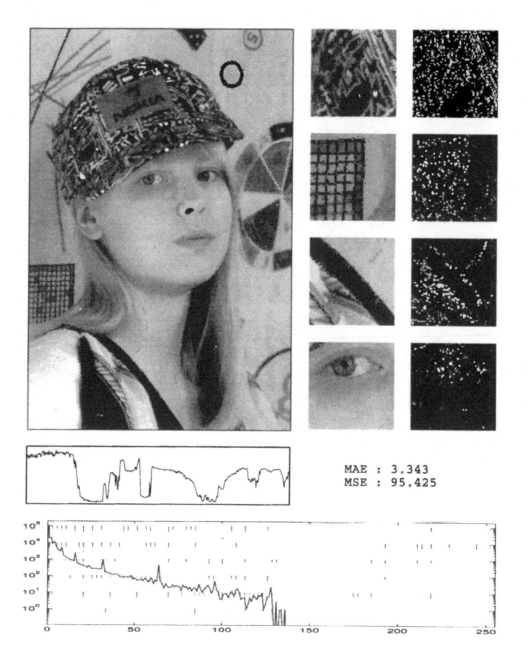

MAE : 3.343
MSE : 95.425

Figure 3.25. "Leena" filtered by the weighted median filter of window size 5×5 and weights a_2 shown in (3.9).

Ranked-order filter

Inputs: *NumberOfRows* × *NumberOfColumns* image
Moving window W, $|W| = N = 2k + 1$
Rank r
Output: *NumberOfRows* × *NumberOfColumns* image

for $i = 1$ to *NumberOfRows*
    for $j = 1$ to *NumberOfColumns*
        place the window W at (i, j)
        store the image values inside W in $\mathbf{x} = (X_1, X_2, \ldots, X_N)$
        find the rth order statistic $X_{(r)}$ of $\mathbf{x}$
        let $Output(i, j) = X_{(r)}$
    end
end

Algorithm 3.9. Algorithm for the ranked-order filter.

Weighted order statistic filters are defined in an analogous manner to the weighted median filter. The output of the rth weighted order statistic filter with weights $\mathbf{a} = (a_1, a_2, \ldots, a_N)$ and input samples $(X_1, X_2, \ldots, X_N)$ is given by

$$\text{WOS}(X_1, X_2, \ldots, X_N; \mathbf{a}, r) = r\text{th order statistic of } \{a_1 \Diamond X_1, a_2 \Diamond X_2, \ldots, a_N \Diamond X_N\}.$$

Weighted order statistic filters can be used to eliminate the negative (positive) impulses and preserve positive (negative) impulses. The weighted order statistic filters can improve the detail preservation properties of the ranked-order filters in a manner similar to the improvement provided by the weighted median filters over the standard median filters. An algorithm for weighted order statistic filtering is shown in Algorithm 3.10.

Example 3.10. Let the 5-point input be $\mathbf{x} = (2, 7, 9, 20, 5)$, the weight vector be $\mathbf{a} = (1, 2, 3, 2, 1)$, and $r = 4$. The output can be found in the following two ways:

- By duplicating the input samples we obtain the multiset $\{2, 7, 7, 9, 9, 9, 20, 20, 5\}$. The ordered multiset now is $\{2, 5, 7, 7, 9, 9, 9, 20, 20\}$. Thus, the 4th order statistic of this multiset and the output is 7.

- The ordered inputs with corresponding weights are $(2, 1), (5, 1), (7, 2), (9, 3), (20, 2)$. The weight of the smallest sample is $1 < 4$, and thus 2 is not the output. The sum of the weights of the two smallest samples equals $1 + 1 < 4$, thus 5 cannot be the output either. Since the sum of the weights of the three smallest samples equals $1 + 1 + 2 \geq 4$, $X_{(3)} = 7$ is the output.

3.6.2 Impulse and Step Response

The impulse response of the rth ranked-order filter is a zero-valued sequence for $r \neq 1, N$. The case $r = 1$ expands negative impulses, whereas the case $r = N$ expands positive impulses. It is easy to verify that the weighted order statistic filters completely remove impulses if each weight a_i satisfies $a_i < \text{MIN}\{r, 1 - r + \sum_{j=1}^{N} a_j\}$ (Exercise).

Weighted order statistic filter

Inputs: $NumberOfRows \times NumberOfColumns$ image
            Moving window W, $|W| = N = 2k + 1$
            Weight vector $a = (a_1, a_2, \ldots, a_N)$
            Natural number r, $1 \leq r \leq \sum_{i=1}^{N} a_i$
Output: $NumberOfRows \times NumberOfColumns$ image

```
for i = 1 to NumberOfRows
    for j = 1 to NumberOfColumns
        place the window W at (i, j)
        store the image values inside W and the corresponding
            weights in x = ((X_1, a_1), (X_2, a_2), ..., (X_N, a_N))
        sort x with respect to X_i s, store the result in
            y = ((X_(1), a_(1)), (X_(2), a_(2)), ..., (X_(N), a_(N)))
        let Sum = 0,  m = 1
        repeat
            let Sum = Sum + a_(m)
            let m = m + 1
        until Sum >= r
        let Output(i, j) = X_(m-1)
    end
end
```

Algorithm 3.10. Algorithm for the weighted order statistic filter.

Ranked-order filters do not blur step edges but translate their position if $r \neq k + 1$. Similarly, weighted order statistic filters cannot blur step edges but may translate their position.

3.6.3   Filtering Examples

We used in the one-dimensional examples $r = 3, 9$ and in the two-dimensional examples $r = 7, 19$ for the ranked-order filters.

For the weighted order statistic filters the weight vector used for the one-dimensional signals is $a_1 = (3, 3, 3, 7, 11, 19, 11, 7, 3, 3, 3)$; in the two-dimensional examples the weights are:

$$
a = \begin{pmatrix} 1 & 1 & 2 & 1 & 1 \\ 1 & 3 & 4 & 3 & 1 \\ 2 & 4 & 11 & 4 & 2 \\ 1 & 3 & 4 & 3 & 1 \\ 1 & 1 & 2 & 1 & 1 \end{pmatrix} . \tag{3.10}
$$

In the one-dimensional example $r = 20$ and in the two-dimensional examples $r = 16$. These numbers are chosen in a way that $r / \sum_{i=1}^{11} a_i \approx 3/11$ ($r / \sum_{i=1}^{25} a_i \approx 9/25$), i.e., they are of the same ratio as was used in the ranked-order filtering. The so obtained improved detail preservation is evident.

The bias is clearly visible in every example and is rather disturbing in some of the examples.

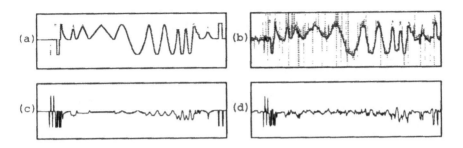

Figure 3.26. One-dimensional signals filtered by the ranked-order filter of window length 11 and $r = 3$.

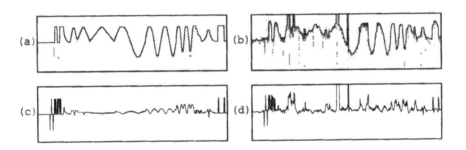

Figure 3.27. One-dimensional signals filtered by the ranked-order filter of window length 11 and $r = 9$.

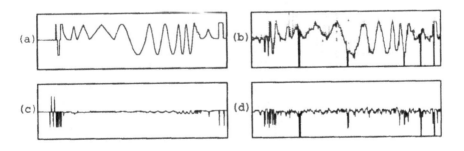

Figure 3.28. One-dimensional signals filtered by the weighted order statistic filter of window length 11, weights $(3, 3, 3, 7, 11, 19, 11, 7, 3, 3, 3)$ and $r = 19$.

Figure 3.29. "Geometrical" filtered by the ranked-order filter of window size 5×5 and $r = 7$.

Figure 3.30. "Geometrical" filtered by the ranked-order filter of window size 5×5 and $r = 19$.

Figure 3.31. "Geometrical" filtered by the weighted order statistic filter of window size 5×5, weights a shown in (3.10), $r = 16$.

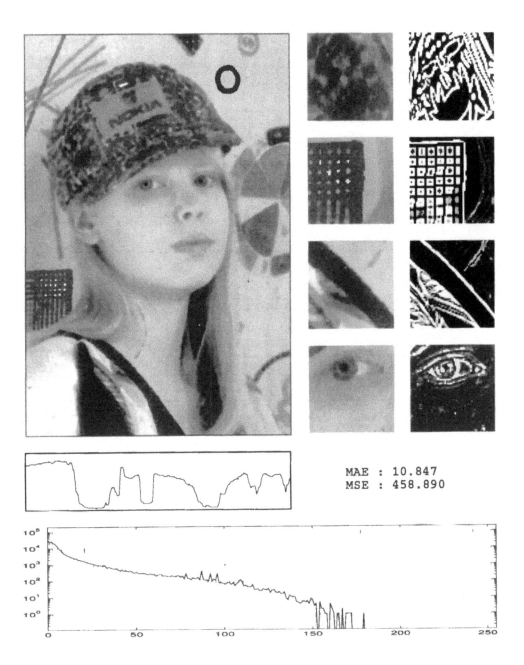

MAE : 10.847
MSE : 458.890

Figure 3.32. "Leena" filtered by the ranked-order filter of window size 5×5 and $r = 7$.

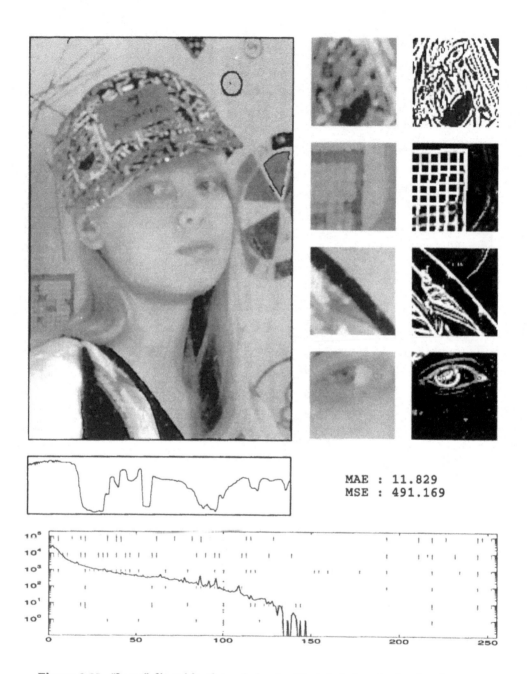

MAE : 11.829
MSE : 491.169

Figure 3.33. "Leena" filtered by the ranked-order filter of window size 5×5 and $r = 19$.

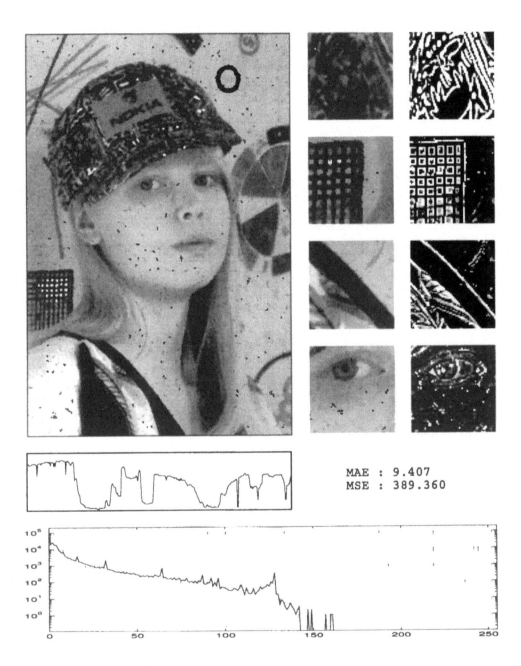

MAE : 9.407
MSE : 389.360

Figure 3.34. "Leena" filtered by the weighted order statistic filter of window size 5×5, weights a shown in (3.10), $r = 16$.

3.7 Multistage Median Filters

3.7.1 Principles and Properties

A problem with the median filter mentioned in Chapter 1 was the calculation of the median when the window size is large. This is often the case in image filtering, where the window size may be up to 9×9. In *multistage median filters*, instead of one median operation, several stages of median filters are used, i.e., the median of medians is used [28, 106]. Another reason for using multistage median filters is to take into account the temporal-order (spatial-order) information better. The third reason, related to the second one, is to obtain filters allowing more design flexibilities than the median filter.

The *separable two-dimensional median filter* [78] is a modification of the median filter yielding filters having properties comparable to the median filters but being much faster. Note that this filter class is not defined for one-dimensional signals.

The separable two-dimensional median filter is obtained by using two successive one-dimensional median filters of window size $z = 2l + 1$ ($z \cdot z = N$), the first one along rows and the second one along the columns of the so-obtained image:

$$
\text{SepMed}\begin{pmatrix} X_{1,1} & X_{1,2} & \ldots & X_{1,z} \\ X_{2,1} & X_{2,2} & \ldots & X_{2,z} \\ \vdots & \vdots & \ddots & \vdots \\ X_{z,1} & X_{z,2} & \ldots & X_{z,z} \end{pmatrix}
$$
$$
= \text{MED}\{\text{MED}\{X_{1,1}, X_{1,2}, \ldots, X_{1,z}\}, \ldots, \text{MED}\{X_{z,1}, X_{z,2}, \ldots, X_{z,z}\}\}.
$$

Example 3.11. Consider the 5×5 image

1	9	12	3	5
11	9	6	6	21
16	5	9	8	8
13	7	12	7	5
22	29	12	15	5

.

Separable two-dimensional median filter

Inputs: *NumberOfRows* × *NumberOfColumns* image
      Moving horizontal window W_1, $|W_1| = z = 2l + 1$
      Moving vertical window W_2, $|W_2| = z = 2l + 1$
Output: *NumberOfRows* × *NumberOfColumns* image

```
median filter the image by using horizontal moving window W₁
store the resulting image in a temporary image
median filter the temporary image by using vertical moving window W₂
```

Algorithm 3.11. Algorithm for the separable two-dimensional median filter.

After filtering the image with the horizontal median filter of length 5 the result is

$$
\begin{array}{|c|}
\hline
5 \\
\hline
9 \\
\hline
8 \\
\hline
7 \\
\hline
15 \\
\hline
\end{array}
$$

When the resulting image is filtered by the vertical median filter of length 5, the result is 8, which is also the output of the separable two-dimensional median filter. Note that this output is not the same as the median of the window. Intuitively, the output of the separable two-dimensional median filter is always close to the median of the window.

In the separable median filter we have two stages of median operations, the first stage in the horizontal direction and the second stage in the vertical. This idea can be easily generalized by using various stages and arbitrary subfiltering operations [80, 106]. Let us denote the results of the horizontal, vertical, two diagonal, cross-shaped, and the X-shaped median filters by (see Figure 3.35)

$$
\begin{aligned}
\text{h-med} \quad &= \quad \text{MED}\{X_{l+1,1}, X_{l+1,2}, \ldots, X_{l+1,z}\}, \\
\text{v-med} \quad &= \quad \text{MED}\{X_{1,l+1}, X_{2,l+1}, \ldots, X_{z,l+1}\}, \\
\text{d45-med} \quad &= \quad \text{MED}\{X_{z,1}, X_{z-1,2}, \ldots, X_{1,z}\}, \\
\text{d135-med} \quad &= \quad \text{MED}\{X_{1,1}, X_{2,2}, \ldots, X_{z,z}\}, \\
\text{c-med} \quad &= \quad \text{MED}\{X_{l+1,1}, X_{l+1,2}, \ldots, X_{l+1,z}\} \cup \{X_{1,l+1}, X_{2,l+1}, \ldots, X_{z,l+1}\}, \\
\text{x-med} \quad &= \quad \text{MED}\{X_{z,1}, X_{z-1,2}, \ldots, X_{1,z}\} \cup \{X_{1,1}, X_{2,2}, \ldots, X_{z,z}\}.
\end{aligned}
$$

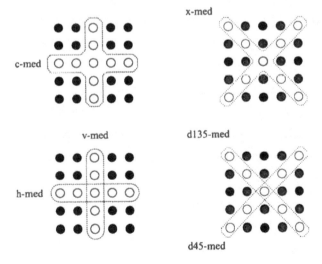

Figure 3.35. Subfilter windows for multistage median filters.

Multistage median filter

Inputs: *NumberOfRows* × *NumberOfColumns* image
            Set of subwindows $\{W_1, W_2, \ldots, W_M\}$
Output: *NumberOfRows* × *NumberOfColumns* image

```
for i = 1 to NumberOfRows
    for j = 1 to NumberOfColumns
        for k = 1 to M
            place the subwindow W_k at (i, j)
            find the median Med(k) of the elements inside W_k
        end
        find the median Med of the vector (Med(1), Med(2), ..., Med(M))
        let Output(i, j) = Med
    end
end
```

Algorithm 3.12. Algorithm for multistage median filtering.

Then, e.g., the following multistage median filters can be obtained:

$$MSM1 = MED\{X_{l+1,l+1}, \text{h-med}, \text{v-med}\},$$
$$MSM2 = MED\{X_{l+1,l+1}, \text{d45-med}, \text{d135-med}\},$$
$$MSM3 = MED\{\text{h-med}, \text{v-med}, \text{d45-med}, \text{d135-med}\},$$

$$MSM4 = MED\{X_{l+1,l+1}, \text{h-med}, \text{v-med}, \text{d45-med}, \text{d135-med}\},$$
$$MSM5 = MED\{X_{l+1,l+1}, MED\{X_{l+1,l+1}, \text{h-med}, \text{v-med}\},$$
$$\qquad\qquad\qquad MED\{X_{l+1,l+1}, \text{d45-med}, \text{d135-med}\}\},$$
$$MSM6 = MED\{X_{l+1,l+1}, \text{c-med}, \text{x-med}\}.$$

The filters MSM1,MSM2,MSM3,MSM4, and MSM5 can be called *unidirectional* because their subfilters utilize median operations along one direction in each of the medians h-med, v-med, d45-med, and d135-med. The filter MSM6 in turn is a *bidirectional* multistage median filter.

Algorithm 3.12 shows how a general multistage filter can be performed. This algorithm is general as the subfilters can be arbitrary as well as their number. Multistage median filters has relations to the *max/median filters* [4] where the output is defined as

$$MaxMed = MAX\{\text{h-med}, \text{v-med}, \text{d45-med}, \text{d135-med}\}.$$

This filter can be used for images in which the geometrical information is on a lower valued background. The max operation introduces a strong bias towards large values. Naturally, one may generalize this filter by using, instead of max or median, some other order statistic or some weighted order statistic.

Example 3.12. Consider the 5 × 5 image

1	9	12	3	5
11	9	6	6	21
16	5	9	8	8
13	7	12	7	5
22	29	12	15	5

Now, we obtain

- h-med = MED$\{16, 5, 9, 8, 8\} = 8$;   • d135-med = MED$\{1, 9, 9, 7, 5\} = 7$;
- v-med = MED$\{12, 6, 9, 12, 12\} = 12$; • c-med = MED$\{16, 5, 9, 8, 8, 12, 6, 12, 12\} = 9$;
- d45-med = MED$\{22, 7, 9, 6, 5\} = 7$;   • x-med = MED$\{22, 7, 9, 6, 5, 1, 9, 7, 5\} = 7$

and so

- MSM1 = MED$\{9, 8, 12\} = 9$;
- MSM2 = MED$\{9, 7, 7\} = 7$;
- MSM3 = MED$\{8, 12, 7, 7\} = 7.5$;
- MSM4 = MED$\{9, 8, 12, 7, 7\} = 8$;
- MSM5 = MED$\{9, \text{MED}\{9, 8, 12\}, \text{MED}\{9, 7, 7\}\} = 9$;
- MSM6 = MED$\{9, 9, 7\} = 9$;
- MaxMed = MAX$\{8, 12, 7, 7\} = 12$.

3.7.2 Impulse and Step Response

The impulse responses of the multistage median filters are all zero-valued sequences since they are based on the median of medians where the first set of medians remove the impulses completely (except the median of size one $X_{l+1, l+1}$). Thus, in the second stage the impulses form a minority which leads to impulse removal. As we have different stages of median operations and the median can preserve step edges, the multistage median filters in general preserve edges. In MSM3 we take the median of four elements, which is, by obeying the conventional definition, the average of the second and third largest elements. This may lead to moderate edge blurring.

3.7.3 Filtering Examples

We show the results only for the separable median, the MSM5, the MSM6, and the max/median filters. It can be seen that numerically the separable median filter gives comparable results to the median filter but it blurs details in a different manner horizontally and vertically. Of these filters the MSM5 preserved details best but left some impulses to the "Leena"-image. The bias of the max/median filter can also be seen.

Figure 3.36. "Geometrical" filtered by the separable median filter of window size 5 × 5.

Figure 3.37. "Geometrical" filtered by the MSM5 filter of window size 5 × 5.

Figure 3.38. "Geometrical" filtered by the MSM6 filter of window size 5 × 5.

Figure 3.39. "Geometrical" filtered by the max/median filter of window size 5 × 5.

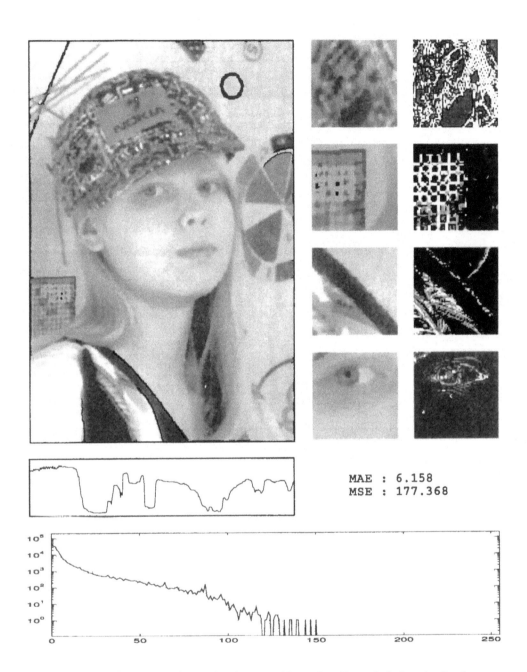

MAE : 6.158
MSE : 177.368

Figure 3.40. "Leena" filtered by the separable median filter of window size 5 × 5.

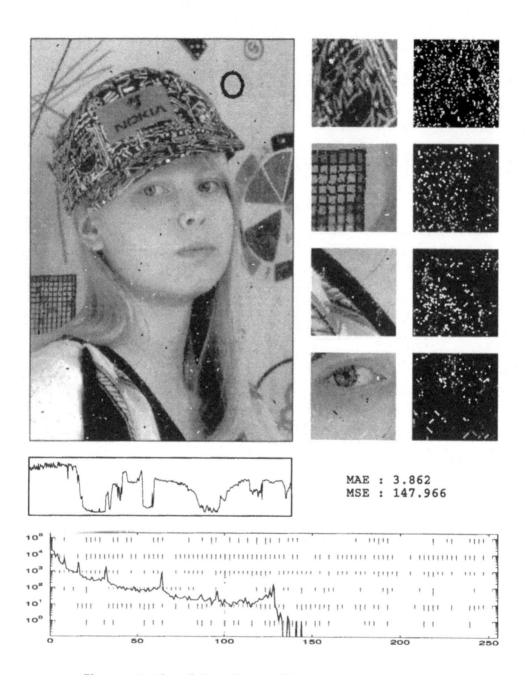

MAE : 3.862
MSE : 147.966

Figure 3.41. "Leena" filtered by the MSM5 filter of window size 5 × 5.

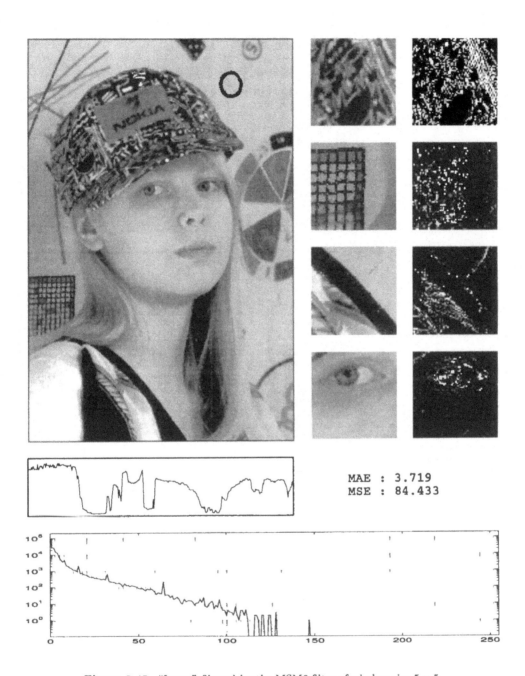

MAE : 3.719
MSE : 84.433

Figure 3.42. "Leena" filtered by the MSM6 filter of window size 5 × 5.

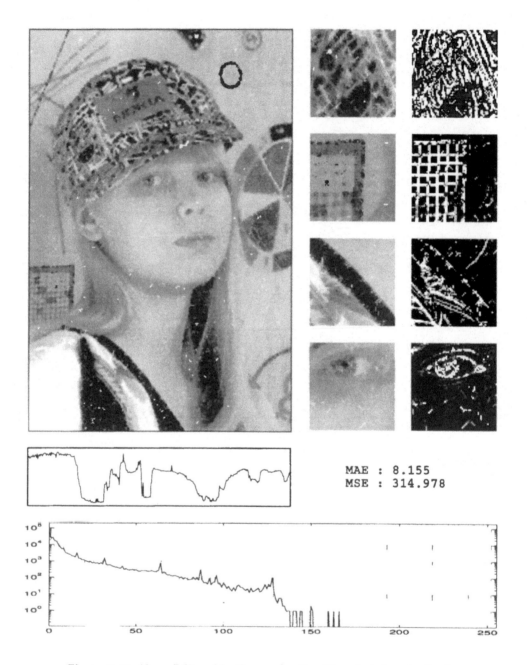

MAE : 8.155
MSE : 314.978

Figure 3.43. "Leena" filtered by the max/median filter of window size 5 × 5.

Median hybrid filter

Inputs: *NumberOfRows* × *NumberOfColumns* image
 Set of subwindows and corresponding weight vectors
 $\{(W_1, a_1), (W_2, a_M) \ldots, (W_z, a_M)\}$
Output: *NumberOfRows* × *NumberOfColumns* image

```
for i = 1 to NumberOfRows
    for j = 1 to NumberOfColumns
        for k = 1 to M
            place the subwindow W_k at (i,j)
            let Mean(k) = the weighted mean of the elements inside W_k
                with weights a_k
        end
        find the median Med of the vector (Mean(1), Mean(2),..., Mean(M))
        let Output(i, j) = Med
    end
end
```

Algorithm 3.13. Algorithm for median hybrid filtering.

3.8 Median Hybrid Filters

3.8.1 Principles and Properties

A combination of linear and nonlinear filters can be obtained by replacing the sub-filters of the multistage median filters by linear filters. This idea has led to the *median hybrid filters* [10, 28, 36]. See Algorithm 3.13. Some of the linear subfilters are typically chosen to follow the slower trends of the input signal and smooth down Gaussian noise. The use of the median facilitates the performance of the filter close to edges and other transition areas. Due to the subfilters, this filter type allows more design possibilities than the median filter. By choosing a quite small number of the linear subfilters, efficient implementation structures can be developed for environments where the compare/swap operations needed to find the median are costly. Median hybrid filters can also be used to alleviate the problem of edge jitter in median filtering.

The output of a median hybrid filter is given by

$$\text{MedHybr}(X_1, X_2, \ldots, X_N) = \text{MED}\{F_1(X_1, X_2, \ldots, X_N), \ldots, F_M(X_1, X_2, \ldots, X_N)\},$$

where the filters $F_1(\cdot), \ldots, F_M(\cdot)$ are linear FIR or IIR filters.

A simple and useful choice for one-dimensional signals is

$$\text{FMH}(X_1, X_2, \ldots, X_N) = \text{MED}\{F_1, F_2, F_3\} = \text{MED}\{(1/k)\sum_{i=1}^{k} X_i, X_{k+1}, (1/k)\sum_{i=k+2}^{N} X_i\}.$$

For this filter, the subfilters $F_1 = (1/k)\sum_{i=1}^{k} X_i$ and $F_3 = (1/k)\sum_{i=k+2}^{N} X_i$ are simple mean filters, the first one is called the backward predictor and the second one the forward predictor. The subfilter $F_2 = X^*$ is an identity filter reacting to all changes in the signal. We call this the basic *FIR median hybrid filter*.

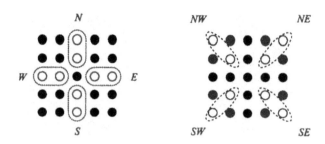

Figure 3.44. Windows for bidirectional median hybrid filters 1LH+ and R1LH+.

Example 3.13. Let the 5-point input be $\mathbf{x} = (2, 7, 9, 20, 5)$ and the subfilters be $F_1 = (X_1 + X_2)/2, F_2 = X_3$, and $F_3 = (X_4 + X_5)/2$. Thus, the outputs of the subfilters are $4.5, 9$ and 12.5 and their median 9 is the output of the basic FIR median hybrid filter.

There exists a large variety of two-dimensional median hybrid filters due to the possibility of choosing the number of subfilters, their type, and also their windows and weights. One example is the so called *1LH+ filter*, [79] defined by

$$1LH+ \begin{pmatrix} X_{1,1} & X_{1,2} & \ldots & X_{1,z} \\ X_{2,1} & X_{2,2} & \ldots & X_{2,z} \\ \vdots & \vdots & \ddots & \vdots \\ X_{z,1} & X_{z,2} & \ldots & X_{z,z} \end{pmatrix}$$

$$= \mathrm{MED}\{Y_N, Y_E, Y_S, Y_W, X_{l+1,l+1}\}$$

$$= \mathrm{MED}\{(1/l)\sum_{i=1}^{l} X_{i,l+1}, (1/l)\sum_{i=l+2}^{z} X_{l+1,i}, (1/l)\sum_{i=l+2}^{z} X_{i,l+1}, (1/l)\sum_{i=1}^{l} X_{l+1,i}, X_{l+1,l+1}\}.$$

Its rotated version is called the *R1LH+ filter*:

$$R1LH+ \begin{pmatrix} X_{1,1} & X_{1,2} & \ldots & X_{1,z} \\ X_{2,1} & X_{2,2} & \ldots & X_{2,z} \\ \vdots & \vdots & \ddots & \vdots \\ X_{z,1} & X_{z,2} & \ldots & X_{z,z} \end{pmatrix}$$

$$= \mathrm{MED}\{Y_{NW}, Y_{SW}, Y_{SE}, Y_{NE}, X_{l+1,l+1}\}$$

$$= \mathrm{MED}\{(1/l)\sum_{i=1}^{l} X_{i,i}, (1/l)\sum_{i=1}^{l} X_{z-i+1,i}, (1/l)\sum_{i=l+2}^{z} X_{i,i}, (1/l)\sum_{i=l+2}^{z} X_{z-i+1,i}, X_{l+1,l+1}\}.$$

The subfilter windows are shown for the 5×5 case in Figure 3.44. The 1LH+ preserves vertical and horizontal lines but not diagonal lines. The R1LH+ preserves diagonal lines but not vertical and horizontal lines. Thus, by adding subfilters with different orientations, the filter becomes less sensitive to the orientation of fine details.

Combining the 1LH+ and R1LH+ filters results in the *2LH+ filter* given by

$$2LH+ \begin{pmatrix} X_{1,1} & X_{1,2} & \ldots & X_{1,z} \\ X_{2,1} & X_{2,2} & \ldots & X_{2,z} \\ \vdots & \vdots & \ddots & \vdots \\ X_{z,1} & X_{z,2} & \ldots & X_{z,z} \end{pmatrix} = \mathrm{MED}\{1LH+, R1LH+, X_{l+1,l+1}\}.$$

This filter is not direction sensitive and it preserves any line in the four basic orientations.

Example 3.14. Consider the 5×5 image

1	9	12	3	5
11	9	6	6	21
16	5	9	8	8
13	7	12	7	5
22	29	12	15	5

Now,

- $Y_N = 0.5(12 + 6) = 9$;
- $Y_E = 0.5(8 + 8) = 8$;
- $Y_S = 0.5(12 + 12) = 12$;
- $Y_W = 0.5(16 + 5) = 10.5$;

- $Y_{NW} = 0.5(1 + 9) = 5$;
- $Y_{SW} = 0.5(22 + 7) = 14.5$;
- $Y_{SE} = 0.5(5 + 7) = 6$;
- $Y_{NE} = 0.5(5 + 6) = 5.5$,

and so the output of the

- 1LH+ filter equals MED$\{9, 8, 12, 10.5, 9\} = 9$;
- R1LH+ filter equals MED$\{5, 14.5, 6, 5.5, 9\} = 6$;
- 2LH+ filter equals MED$\{9, 6, 9\} = 9$.

The median can naturally be replaced by any other order statistic or, more generally, any weighted order statistic. Then the *FIR-WOS hybrid filters* are obtained [118, 119].

On stationary areas of the signal corrupted by Gaussian noise, the noise attenuation can be improved by making longer averaging subwindows as the output variance of the subwindows is inversely proportional to their lengths. However, in the vicinity of a noisy step edge the performance of the FIR median hybrid filter with long averages deteriorates. Thus, there is a need to use different subwindow lengths when filtering a nonstationary signal. *In-place growing FIR median hybrid filters* offer an efficient solution for combining the FIR median hybrid filters of different lengths [114]. The filter window is divided into centered and overlapping FIR median hybrid filters of different lengths. The output of the shortest filter replaces the center point of the next longest filter and so forth until the procedure has advanced to the final stage with the longest averaging subfilters. The structure with $2M$ averaging subfilters is defined recursively as

$$Y_0 = X_{k+1}$$

$$Y_j = \text{MED}\{(1/k_j)\sum_{i=1}^{k_j} X_{k+1-i}, Y_{j-1}, (1/k_j)\sum_{i=1}^{k_j} X_{k+1+i}\}$$

$$\text{IPGFMH}(X_1, X_2, \ldots, X_N) = Y_M,$$

where $k_j > k_{j-1}$ and $j = 1, 2, \ldots, M$. This can be easily generalized by having any number of subfilters at every stage instead of two.

Example 3.15. Let the 7-point input be $\mathbf{x} = (4, 2, 7, 9, 20, 5, 1)$ and the lengths of the subfilters be 2 and 3. Thus, the output of the first stage is MED$\{(2+7)/2, 9, (20+5)/2\} = 9$ and the output of the second and the final stage is MED$\{(4+2+7)/3, 9, (20+5+1)/3\} = 26/3$.

This concept can be easily generalized for two-dimensional filters as well by using median hybrid filters with in-place growing FIR median hybrid subfilters [115].

3.8.2   Impulse and Step Response

It is obvious that the basic median hybrid filters remove impulses as the effect of the impulse can be in only one of the subfilters, thus not affecting the result of the median operation. Furthermore, the basic median hybrid filters preserve step edges. This is seen as follows: The subfilter F_1 leads the edge, F_3 lags the edge, and F_2 reacts at the edge. Thus, their median follows the edge. The same also holds for the in-place growing FIR median hybrid filters. The other median hybrid filters remove impulses and preserve edges as well.

3.8.3   Filtering Examples

In these examples the basic FIR median hybrid filter and the in-place growing FIR median hybrid filter behaved almost equivalently.

From the filtered "Geometrical" image we see how the 1LH+ works better for structures in horizontal and vertical orientations than the R1LH+, whereas for diagonal orientations the order is the opposite one. Furthermore, it is seen how the 2LH+ successfully combines good points of the 1LH+ and the R1LH+. Clearly, the 1LH+ worked much better with the "Leena" image.

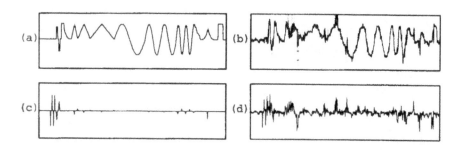

Figure 3.45. One-dimensional signals filtered by the basic FIR median hybrid filter of window length 11.

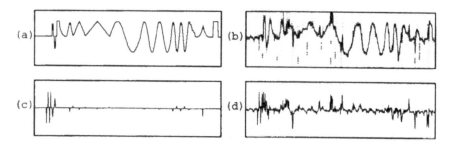

Figure 3.46. One-dimensional signals filtered by the in-place growing FIR median hybrid filter of window length 11, subwindow lengths 2, 3, 4, and 5.

Figure 3.47. "Geometrical" filtered by the 1LH+ filter of window size 5 × 5.

Figure 3.48. "Geometrical" filtered by the R1LH+ filter of window size 5 × 5.

Figure 3.49. "Geometrical" filtered by the 2LH+ filter of window size 5 × 5.

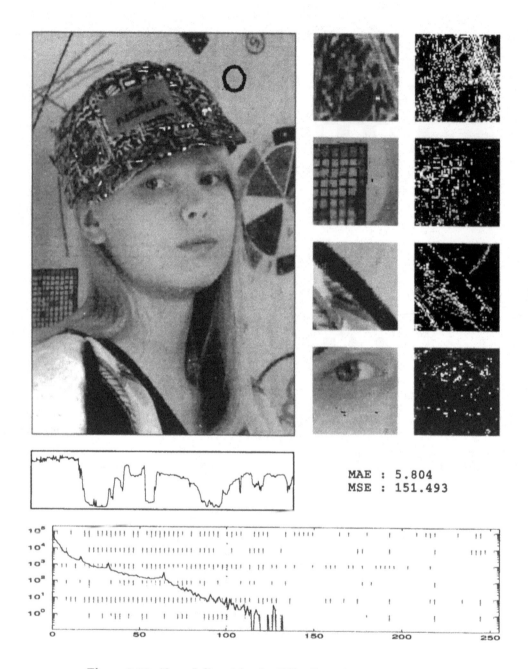

MAE : 5.804
MSE : 151.493

Figure 3.50. "Leena" filtered by the 1LH+ filter of window size 5 × 5.

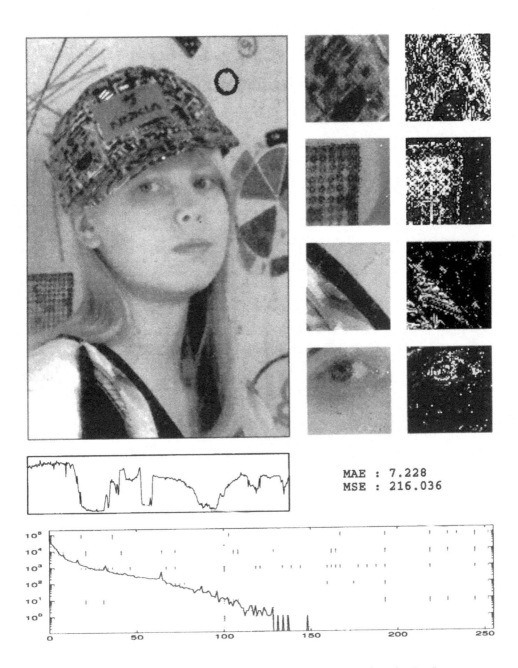

MAE : 7.228
MSE : 216.036

Figure 3.51. "Leena" filtered by the R1LH+ filter of window size 5 × 5.

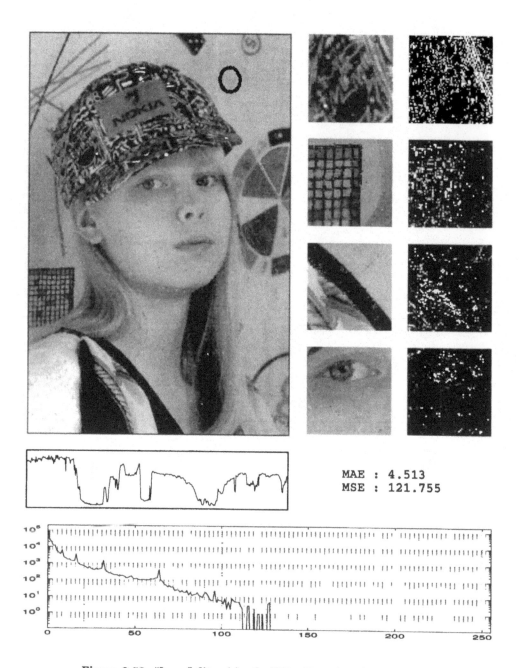

MAE : 4.513
MSE : 121.755

Figure 3.52. "Leena" filtered by the 2LH+ filter of window size 5 × 5.

Hachimura-Kuwahara filter

Inputs: *NumberOfRows* × *NumberOfColumns* image
Set of subwindows $\{W_1, W_2, \ldots, W_M\}$
Output: *NumberOfRows* × *NumberOfColumns* image

```
for i = 1 to NumberOfRows
    for j = 1 to NumberOfColumns
        let SmallestVar = LargestPossibleValue
        for k = 1 to M
            place the subwindow Wₖ at (i, j)
            let Varₖ = the variance of the elements inside Wₖ
            if Varₖ < SmallestVar
                let SmallestVar = Varₖ
                let Meanₖ = the mean of the elements inside Wₖ
                let OutputCandidate = Meanₖ
        end
        let Output(i, j) = OutputCandidate
    end
end
```

Algorithm 3.14. Algorithm the for Hachimura-Kuwahara filter.

3.9 Edge-Enhancing Selective Filters

3.9.1 Principles and Properties

Selective filters are multilevel filters where several subwindows are used, and based on some criterion the output is chosen to be one of the outputs of these windows. Median hybrid filters are examples of these selective filters, where the selection rule is given in the form of the median of the outputs of the subfilters. The objective of using a selective scheme is often to obtain edge-enhancing filters. We have already seen that mean and median filters cannot enhance blurred edges, e.g., they do not change ramps into more step-like structures. In fact the median filter preserves any monotonic degradation of the edge (Exercise).

In Reference [57] an edge-enhancing smoother was introduced. This smoother is called the *Hachimura-Kuwahara filter*, where the means and variances of the subwindows are calculated and the output is given by the mean from the window having the smallest variance; see Algorithm 3.14. Thus, the idea is to take the output from the most homogeneous neighborhood area. Some tie-breaking rule must naturally be applied if two (or more) areas have the minimum variance. One may then, e.g., choose the area whose mean value is closest to the value of X^*. Figure 3.53 illustrates Hachimura-Kuwahara filtering of ramp edges. While filtering the ramp the output will be chosen from one of the homogeneous regions before and after the ramp. This leads to edge enhancement.

The article in Reference [57] is not the only article where the same idea has been proposed, but it is the first one we know of. Different proposals typically differ in the numbers and the shapes of the subwindows. In our examples, we used the subwindows shown in Figure 3.54. Since the output is now calculated from a smaller area than the

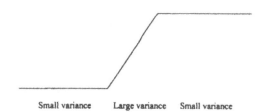

Small variance   Large variance   Small variance

Figure 3.53. A ramp signal and areas with different variance levels.

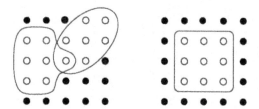

Figure 3.54. Subwindows used in the Hachimura-Kuwahara filter. The rest of the subwindows are obtained by rotating these subwindows $45°, 90°$, and $135°$.

whole window, the noise attenuation capability of the Hachimura-Kuwahara filter is worse than that of the mean filter.

One might as well take some other estimators of scale and location within the subwindows instead of the variance and the mean. For instance, the MAD and the median provide improved robustness in the presence of impulsive noise.

Example 3.16.   Consider the 5×5 image

1	9	12	3	5
11	9	6	6	21
16	5	9	8	8
13	7	12	7	5
22	29	12	15	5

.

Now, by using the subwindows shown in Figure 3.54 we obtain the variances and means:

- $\mu_1 = (1 + 9 + 11 + 9 + 6 + 5 + 9)/7 = 50/7 \approx 7.143, \sigma_1^2 = [(1 - \mu_1)^2 + (9 - \mu_1)^2 + (11 - \mu_1)^2 + (9 - \mu_1)^2 + (6 - \mu_1)^2 + (5 - \mu_1)^2 + (9 - \mu_1)^2]/7 = 482/49 \approx 9.84;$

- $\mu_2 = (9 + 12 + 3 + 9 + 6 + 6 + 9)/7 = 54/7 \approx 7.714, \sigma_2^2 = [(9 - \mu_2)^2 + (12 - \mu_2)^2 + (3 - \mu_2)^2 + (9 - \mu_2)^2 + (6 - \mu_2)^2 + (6 - \mu_2)^2 + (9 - \mu_2)^2]/7 = 360/49 \approx 7.35;$

- $\mu_3 = (3 + 5 + 6 + 6 + 21 + 9 + 8)/7 = 58/7 \approx 8.286, \sigma_3^2 = [(3 - \mu_3)^2 + (5 - \mu_3)^2 + (6 - \mu_3)^2 + (6 - \mu_3)^2 + (21 - \mu_3)^2 + (9 - \mu_3)^2 + (8 - \mu_2)^2]/7 = 1480/49 \approx 30.20;$

  $\vdots$

- $\mu_9 = (9+6+6+5+9+8+7+12+7)/9 = 69/9 \approx 7.667, \sigma_9^2 = [(9-\mu_9)^2+(6-\mu_9)^2+(6-\mu_9)^2+(5-\mu_9)^2+(9-\mu_9)^2+(8-\mu_9)^2+(7-\mu_9)^2+(12-\mu_9)^2+(7-\mu_9)^2]/9 = 4;$

The smallest variance was in the 3×3 square subwindow W_9, whose mean, $69/9$, is the output.

Figure 3.55. (a) A ramp edge. (b) Results after mean and median filtering.

Another edge-enhancing smoother, called the *comparison and selection filter*, is introduced in Reference [65]. This filter is based on the observation that the output of the median filter is smaller than the output of the mean filter before a rising edge; and the output of the median filter is larger than the output of the mean filter after a rising edge; see Figure 3.55. Thus, in order to enhance a blurred edge, the value of some smaller sample than the median (for example the minimum sample) is chosen for the output if the median is smaller than the mean; otherwise the value of some larger sample than the median (for example the maximum sample) is chosen for the output. More formally, let $1 \leq j \leq k$, then

$$\mathrm{CS}(X_1, X_2, \ldots, X_N; j) = \begin{cases} X_{(k+1-j)}, \mathrm{MEAN}\{X_1, X_2, \ldots, X_N\} \geq \mathrm{MED}\{X_1, X_2, \ldots, X_N\}, \\ X_{(k+1+j)}, \text{otherwise.} \end{cases}$$

Algorithm 3.15 gives a routine for implementing the comparison and selection filter. Again the edges are enhanced at the expense of noise suppression.

Comparison and selection filter

Inputs: *NumberOfRows* × *NumberOfColumns* image
Moving window $W, |W| = N = 2k + 1$
Natural number j, $1 \leq j \leq k$
Output: *NumberOfRows* × *NumberOfColumns* image

```
for i = 1 to NumberOfRows
    for j = 1 to NumberOfColumns
        place the window W at (i, j)
        find the median Med and the mean Mean of the values inside W
        if Mean ≥ Med
            find the (k + 1 − j)th smallest value X_(k+1−j) inside W
            let Output(i, j) = X_(k+1−j)
        else
            find the (k + 1 + j)th smallest value X_(k+1+j) inside W
            let Output(i, j) = X_(k+1+j)
        end
    end
end
```

Algorithm 3.15. Algorithm for the comparison and selection filter.

Example 3.17. Let the 7-point input be $\mathbf{x} = (2, 2, 4, 6, 8, 10, 10)$, the window length be 5 and $j = 1$. The mean and the median of the first window position are $(2+2+4+6+8)/5 = 4.4$ and 4, respectively. As the mean is larger, the output is $X_{(k+1-j)} = X_{(2+1-1)} = X_{(2)} = 2$. In the next window position we have $mean = 6 = median$, thus the output is $X_{(2)} = 4$. In the third window position we obtain: $mean = 7.6 < med = 8$, thus the output is $X_{(4)} = 10$. Clearly, the edge is enhanced by the filter.

The idea behind the *selective average filter* [49] and the *selective median filter* [49] is closely related to that of median hybrid filters. Similarly, two predictors, a backward predictor and a forward predictor, are used. The predictors are the mean (median) of the k previous samples and the mean (median) of the k following samples for the selective average (median) filter. Instead of taking the median of the results of these predictors and the middle sample $X^* = X_{k+1}$, the mean (median) which is closest to X_{k+1} is the output; see Algorithm 3.16. It is easy to understand that this will lead to enhancement of blurred step edges. Let us denote by $mean_b = \text{MEAN}\{X_1, X_2, \ldots, X_k\}$, $mean_f = \text{MEAN}\{X_{k+2}, X_{k+3}, \ldots, X_N\}$, $med_b = \text{MED}\{X_1, X_2, \ldots, X_k\}$, and $med_f = \text{MED}\{X_{k+2}, X_{k+3}, \ldots, X_N\}$. The outputs of the selective average and selective median filters are then given by

$$\text{SelAve}(X_1, X_2, \ldots, X_N) = \begin{cases} mean_b, & \text{if } |mean_b - X_{k+1}| \leq |mean_f - X_{k+1}|, \\ mean_f, & \text{otherwise} \end{cases}$$

and

$$\text{SelMed}(X_1, X_2, \ldots, X_N) = \begin{cases} med_b, & \text{if } |med_b - X_{k+1}| \leq |med_f - X_{k+1}|, \\ med_f, & \text{otherwise}. \end{cases}$$

These filters can simultaneously enhance blurred edges and suppress noise rather efficiently. Furthermore, they are computationally more efficient than the Hachimura-Kuwahara filter since the sample variances are not evaluated. The two-dimensional version of the selective average (median) filter can also be defined, but it is no longer as natural as the one-dimensional version because of the wide variety of possible edge orientations.

Selective average filter

```
Inputs: Signal of length Length
        Moving window W, |W| = N = 2k + 1
Output: Signal of length Length

for i = 1 to Length
    find the mean Mean1 of the k previous samples
    find the mean Mean2 of the k following samples
    if |Mean1 - X*| ≤ |Mean2 - X*|
        let Output(i) = Mean1
    else
        let Output(i) = Mean2
end
```

Algorithm 3.16. Algorithm for the selective average filter.

Example 3.18. Let the 7-point input be $x = (2, 2, 4, 6, 8, 10, 10)$, the window length be 5 and $j = 1$. Consider first the selective average filter. The means in the first window position are $(2 + 2)/2 = 2$ and $(6 + 8)/2 = 7$, of which 2 is closer to 4, and thus it is the output. In the next window position $mean_b = 3$ and $mean_f = 9$. Thus, the output is 3. In the last window position the means are $mean_b = 5$ and $mean_f = 10$, and thus the output is 10. As the window size for the predictors is two, the mean and the median coincide. Thus, the obtained output sequence is also the output sequence for the selective median filter. These filters enhanced the edge as expected.

3.9.2 Impulse and Step Response

Since an impulse always gives high increase to the variance, the impulses are removed by the Hachimura-Kuwahara filter. The comparison and selection filter removes impulses for every j (Exercise). It is also easy to show that the selective average (median) filter removes impulses (Exercise). Furthermore, the Hachimura-Kuwahara filter, the comparison and selection filter, and the selective average (median) filter preserve step edges (Exercise).

3.9.3 Filtering Examples

The subwindows used for the Hachimura-Kuwahara filter in the one-dimensional case are $W_1 = \{1, 2, 3, 4\}, W_2 = \{5, 6, 7\}, W_3 = \{8, 9, 10, 11\}$, and in the two-dimensional case they are shown in Figure 3.54.

It can be seen that these filters enhance edges but at the same time create some undesired artifacts. Furthermore, they are not very efficient in noise cancellation.

From Figure 3.62 we can also see that in the case of impulses they dominate the variance, and the most homogeneous area can be in almost any of the subwindows. Because the mean is also sensitive to the impulses the result is pretty granular.

3.10 Rank Selection Filters

3.10.1 Principles and Properties

Rank selection filters fall in the general framework of selection filters [32]. A block diagram illustrating rank selection filtering is shown in Figure 3.64. It consists of three operative different blocks. One of the blocks extracts some information from the samples $x = (X_1, X_2, \ldots, X_N)$. This information is stored into the feature vector z. The second block produces the sorted vector $x_{()}$. Finally, the third block, the output rank selector, chooses the most appropriate sample from $x_{()} = (X_{(1)}, X_{(2)}, \ldots, X_{(N)})$ to be the output. This choice is based on the feature vector z and the rank selection rule $S(\cdot)$

The feature vector represents part of the information contained in x. In general, the full information contained in the observation vector x is not used as the basis for determining the output. On the other hand, the performance of the rank selection filters depends heavily on the choice of the information used to form the feature vector. Therefore, it is necessary to extract from x the information which is the most relevant for the application in hand. For example, the comparison and selection filter belongs to the framework of rank selection filtering. There the extracted information is in the form of two real values: the mean and the median of the samples. Based on their ordering, the output sample is chosen from the set $\{X_{(k+1-j)}, X_{(k+1+j)}\}$.

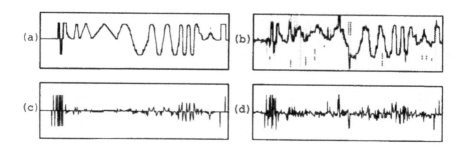

Figure 3.56. One-dimensional signals filtered by the Hachimura-Kuwahara filter of window length 11.

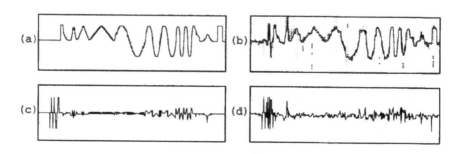

Figure 3.57. One-dimensional signals filtered by the comparison and selection filter of window length 11, $j = 2$.

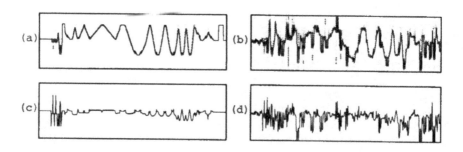

Figure 3.58. One-dimensional signals filtered by the selective average filter of window length 11.

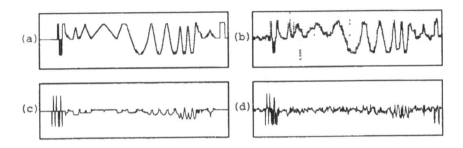

Figure 3.59. One-dimensional signals filtered by the selective median filter of window length 11.

Figure 3.60. "Geometrical" filtered by the Hachimura-Kuwahara filter of window size 5×5.

Figure 3.61. "Geometrical" filtered by the comparison and selection filter of window size 5×5, $j = 2$.

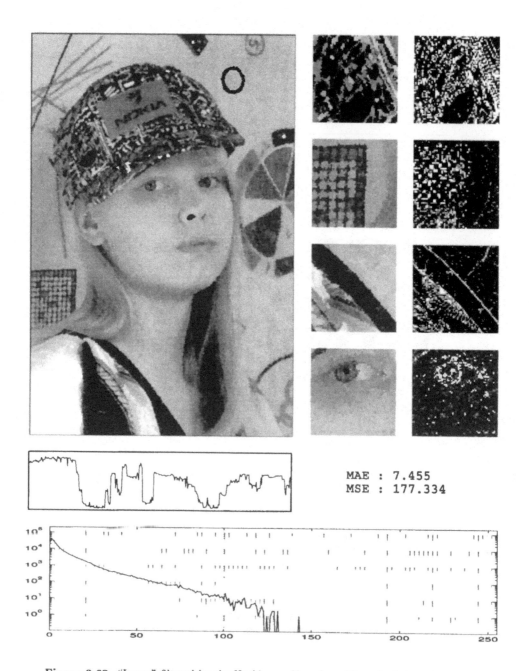

MAE : 7.455
MSE : 177.334

Figure 3.62. "Leena" filtered by the Hachimura-Kuwahara filter of window size 5 × 5.

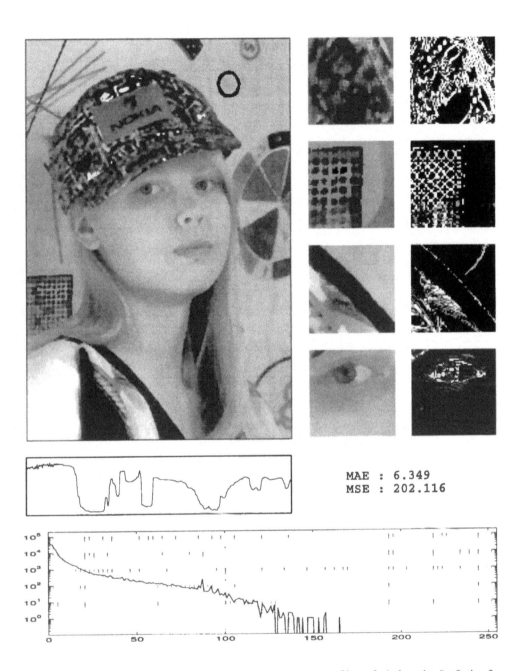

MAE : 6.349
MSE : 202.116

Figure 3.63. "Leena" filtered by the comparison and selection filter of window size 5×5, $j = 2$.

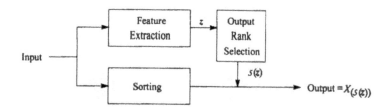

Figure 3.64. Rank selection filtering structure (adapted from Reference [32]).

Another way to understand the rank selection filters is to consider $S(\cdot)$ as a classifier partitioning the feature space where the feature vectors are lying into N regions. Whenever the formed feature vector lies in the ith partition, the filter output is $X_{(i)}$, the ith order statistic.

In addition to the comparison and selection filter, some of the above-mentioned filters can be formulated as rank selection filters. For example, perhaps surprisingly, ranked-order statistic filters are such filters. There the feature space is the empty space, i.e., the length of the feature vector is 0, and the selection rule is a constant: $S(\cdot) = i$. However, this interpretation seems quite artificial.

    Permutation filters belong to the rank selection framework [11]. The filter definition follows from the fact that the mapping from the temporally ordered input vector $\mathbf{x}$ to the rank-ordered vector defines a permutation of the elements comprising $\mathbf{x}$. A permutation filter associates with each such permutation a specific order statistic. More formally, let

$$\mathbf{x}^r = (R(X_1), R(X_2), \ldots, R(X_N)),$$

where $R(X_i)$ is the rank of the sample X_i. This vector is called the *sample permutation vector* of $\mathbf{x}$. Clearly, sample permutation vectors belong to the feature space

$$Z = \{(r_1, r_2, \ldots, r_N) : r_i \in \{1, 2, \ldots, N\} \text{ and } r_i \neq r_j \text{ for all } i \neq j\}.$$

The output of the permutation filter defined by the rank selection rule $S : Z \mapsto \{1, 2, \ldots, N\}$ is given by (see Algorithm 3.17)

$$\text{Perm}(X_1, X_2, \ldots, X_N; S) = X_{(S(\mathbf{x}^r))}.$$

Example 3.19. Let the input be $(2, 7, 9, 20, 5)$ and the rank selection rule be given by

$$S(R(X_1), R(X_2), R(X_3)) = \begin{cases} 2, & \text{if } (R(X_1), R(X_2), R(X_3)) = (1,2,3), \\ 3, & \text{if } (R(X_1), R(X_2), R(X_3)) = (1,3,2), \\ 2, & \text{if } (R(X_1), R(X_2), R(X_3)) = (2,1,3), \\ 3, & \text{if } (R(X_1), R(X_2), R(X_3)) = (2,3,1), \\ 1, & \text{if } (R(X_1), R(X_2), R(X_3)) = (3,1,2), \\ 1, & \text{if } (R(X_1), R(X_2), R(X_3)) = (3,2,1). \end{cases}$$

- The first three samples are in an increasing order giving the sample permutation vector $(1,2,3)$, and the result of the output rule for this feature vector is 2. This means that the output is $X_{(2)}$, which is 7 taken from $\{2,7,9\}$.

- Similarly the output for the next three samples is their median $X_{(2)}$, i.e., 9.

- The sample permutation vector for the last three samples equals $(2,3,1)$. The result of the output rule for this feature vector is 3, meaning that the output is the largest value $X_{(3)} = 20$.

Permutation filter

Inputs: *NumberOfRows* × *NumberOfColumns* image
Moving window W, $|W| = N = 2k + 1$
Selection rule $S(\cdot)$
Output: *NumberOfRows* × *NumberOfColumns* image

```
for i = 1 to NumberOfRows
    for j = 1 to NumberOfColumns
        place the window W at (i,j)
        store the image values inside W in x = (X₁, X₂, ..., Xₙ)
        sort x, store the result in y = (X₍₁₎, X₍₂₎, ..., X₍ₙ₎)
        for every element Xᵢ of x find its position R(Xᵢ) in y,
            store the result in r = (R(X₁), R(X₂), ..., R(Xₙ))
        let OutputRank = S(R(X₁), R(X₂), ..., R(Xₙ))
        let Output(i,j) = X₍OutputRank₎
    end
end
```

Algorithm 3.17. Algorithm for the permutation filter.

Notice that one could as well define the permutation filter as associating every permutation with a specific input sample. The idea of permutation filtering is quite obvious and it can be found in literature before the article in Reference [11]. For instance, in Reference [102] the permutation filter is called the *variable rank order statistics filter*. An even more general concept where the permutation filter belongs to can be found in Reference [6].

The main problem with permutation filters is their implementational limits. The number of the permutations for the window size N is $N!$, which is far too large for practical window sizes.

The class of *Rank Conditioned Rank Selection (RCRS) filters* is another class of rank selection filters [32]. This class is intermediate between the rank-order filters (0th order RCRS filter) and the permutation filters (Nth order RCRS filter). The idea is not to take every sample rank into consideration but a subset of the sample ranks. Thus, two permutations are considered to be equivalent if their chosen sample ranks match. In RCRS filtering the output rank is the same for equivalent permutations. The feature vector $\mathbf{z}$ for an RCRS filter consists of the ranks of selected samples in the observation vector $\mathbf{x}$. The selected samples are placed in a vector, $\mathbf{x}^* = (X_{\gamma_1}, X_{\gamma_2}, \ldots, X_{\gamma_M})$, where M, $0 \leq M \leq N$, the amount of selected samples, is called the order of the RCRS filter. The indices $\gamma_1, \gamma_2, \ldots, \gamma_M$ are called the indices of selected samples. We denote the set of the selected indices by $\Gamma = \{\gamma_1, \gamma_2, \ldots, \gamma_M\}$. The respective ranks of these samples form the feature vector, $\mathbf{z} = \mathbf{x}^{r*} = (R(X_{\gamma_1}), R(X_{\gamma_2}), \ldots, R(X_{\gamma_M}))$. The feature space contains all combinations of ranks of M samples excluding those in which two are equal, since those combinations cannot occur. We denote $Z = \{(r_1, r_2, \ldots, r_M) : r_i \in \{1, 2, \ldots, N\}$ and $r_i \neq r_j$ for all $i \neq j\}$.

Let $S : Z \mapsto \{1, 2, \ldots, N\}$. The output of an Mth order RCRS filter of window size N is given by (see Algorithm 3.18)

$$\text{RCRS}(X_1, X_2, \ldots, X_N; S) = X_{(S(\mathbf{x}^{r*}))}.$$

Rank Conditioned Rank Selection filter

Inputs: *NumberOfRows* × *NumberOfColumns* image
       Moving window $W, |W| = N = 2k + 1$
       Subwindow for the feature vector W_{sub}, $W_{sub} \subseteq W$, $|W_{sub}| = M$
       Selection rule S
Output: *NumberOfRows* × *NumberOfColumns* image

```
for i = 1 to NumberOfRows
    for j = 1 to NumberOfColumns
        place the windows W and W_sub at (i,j)
        store the image values inside W in x = (X_1, X_2, ..., X_N)
        store the image values inside W_sub in z = (X_γ1, X_γ2, ..., X_γM)
        sort x, store the result in y = (X_(1), X_(2), ..., X_(N))
        for every element X_γi of z find its position R(X_i) in y,
            store the result in r = (R(X_γ1), R(X_γ2), ..., R(X_γM))
        let OutputRank = S(R(X_γ1), R(X_γ2), ..., R(X_γM))
        let Output(i,j) = X_(OutputRank)
    end
end
```

Algorithm 3.18. Algorithm for the RCRS filter.

The order of the filter and the location of the samples selected for $\mathbf{x}^*$ depend on the application. The larger M is, the more rank information is included into the feature vector and the better filter performance is achieved. On the other hand, a larger M results in an increased complexity in the filter implementation.

Example 3.20. Let the input be $(X_1, X_2, X_3, X_4, X_5) = (2, 7, 9, 20, 5)$ and $\Gamma = \{3\}$. Thus, $M = 1$ and the rank of the sample X_3 forms the feature vector. Let the rank selection rule be given by

$$S(R(X_3)) = \begin{cases} R(X_3), & \text{if } R(X_3) = 2, 3, 4, \\ 3, & \text{otherwise.} \end{cases}$$

Therefore, the output is the center sample if it is not the smallest nor the largest among the samples. Otherwise, the filter outputs the median. In this case, since 9 is not an extreme order statistic, the output is 9.

The filter in Example 3.20 is called the *Rank Conditioned Median* in Reference [32]. Clearly, it rejects impulses by replacing them by the median, and, if the center sample lies in the middle ranks, it is unaltered. This leads to improved detail preservation in comparison to the median filter.

Recently permutation filters and RCRS filter classes were enlarged to form *Extended Permutation Rank Selection filter* classes [33]. In addition to the N observation samples, the extended vector contains K statistics, which are functions of the observation samples. The extended filters are then defined for the new vector in an analogous manner to the permutation filters and the RCRS filters.

LUM filter

Inputs: *NumberOfRows* × *NumberOfColumns* image
        Moving window $W, |W| = N = 2k + 1$
        Natural numbers $s, t, \ 1 \leq s \leq t \leq (N + 1)/2$
Output: *NumberOfRows* × *NumberOfColumns* image

for $i = 1$ to *NumberOfRows*
    for $j = 1$ to *NumberOfColumns*
        place the window W at (i, j)
        store the image values inside W in $\mathbf{x} = (X_1, X_2, \ldots, X_N)$
        sort $\mathbf{x}$, store the result in $\mathbf{y} = (X_{(1)}, X_{(2)}, \ldots, X_{(N)})$
        let *LowerSamples* $= (X_{(s)}, X^*, X_{(t)})$
        let *UpperSamples* $= (X_{(N-t+1)}, X^*, X_{(N-s+1)})$
        find the medians *Upper* and *Lower* of the vectors
            UpperSamples and *LowerSamples*
        if $X^* \leq (Upper + Lower)/2$
            let *Output*$(i, j) = Lower$
        else
            let *Output*$(i, j) = Upper$
        end
    end
end

Algorithm 3.19. Algorithm for the LUM filter.

One advantage of the (extended) permutation and RCRS filters is that they are easy to optimize if a training pair of a noisy and a desired signal is available [32]. We hope that the reader will notice that the nature of this kind of optimization is very different from the one mentioned in Section 3.3 and which will be discussed thoroughly in Chapter 4. There we have some model for the signal (e.g., a constant one) and some model for the noise (e.g., its distribution) and the objective is to find the filter coefficients minimizing some distortion measure. In the training based optimization, instead of models we are using a pair of realizations of the signal and the noise, and are again aiming to find the filter coefficients minimizing some distortion measure.

Lower-Upper-Middle (LUM) filters are also closely related to the RCRS filters [31]. The output of the so called *LUM smoother* with parameter $s, 1 \leq s \leq k + 1$, is given by

$$\text{LUMSmooth}(X_1, X_2, \ldots, X_N; s) = \text{MED}\{X_{(s)}, X^*, X_{(N-s+1)}\}.$$

Thus, the output of the LUM smoother is $X_{(s)}$ (some lower order statistic) if the center sample satisfies $X^* < X_{(s)}$. If $X^* > X_{(N-s+1)}$ the output of the LUM smoother is $X_{(N-s+1)}$ (some higher order statistic). Otherwise the output is simply the center sample, X^*. The parameter s defines the range of the accepted order statistics. If X^* lies in the range it is not modified. If X^* lies outside this range, it is replaced by a sample lying closer to the median. Impulses are typically such values. The parameter value s enables tuning between detail preservation and noise suppression. If $s = 1$ the LUM smoother is an identity filter; if $s = k + 1$ it is the median filter. In fact the LUM smoother is identical to the *Winsorized smoother* proposed by Mallows [73, 74]. Furthermore, it is identical to the center weighted median filter (Exercise).

The *LUM sharpener* moves samples away from the median to more extreme order statistics to obtain sharpening and edge-enhancing characteristics. The sharpening is controlled by the parameter $t, 1 \leq t \leq k+1$. The *midpoint* between two order statistics $X_{(t)}$ (lower) and $X_{(N-t+1)}$ (upper) is given by

$$t_l = \frac{X_{(t)} + X_{(N-t+1)}}{2}.$$

The output of the LUM sharpener with parameter t is given by

$$\text{LUMSharp}(X_1, X_2, \ldots, X_N; t) = \begin{cases} X_{(t)}, & \text{if } X_{(t)} < X^* \leq t_l, \\ X_{(N-t+1)}, & \text{if } t_l < X^* \leq X_{(N-t+1)}, \\ X^*, & \text{otherwise.} \end{cases}$$

If $t = k+1$ the LUM sharpener is simply the identity filter. If $t = 1$ a maximum amount of sharpening is obtained since X^* is being shifted to one of the extreme order statistics $X_{(1)}$ or $X_{(N)}$. In case of a tie it must be broken some way.

When the philosophies of the LUM smoother and the LUM sharpener are combined, the *LUM filter* is obtained. The output of the LUM filter with parameters s and t, $1 \leq s \leq t \leq k+1$, is given by

$$\text{LUM}(X_1, X_2, \ldots, X_N; s, t) = \begin{cases} X_{(s)}, & \text{if } X^* < X_{(s)}, \\ X_{(t)}, & \text{if } X_{(t)} < X^* \leq t_l, \\ X_{(N-t+1)}, & \text{if } t_l < X^* \leq X_{(N-t+1)}, \\ X_{(N-s+1)}, & \text{if } X_{(N-s+1)} < X^*, \\ X^*, & \text{otherwise.} \end{cases} \qquad (3.11)$$

Thus, extreme samples, $X^* < X_{(s)}$ or $X^* > X_{(N-s+1)}$, are smoothed by shifting them towards the median, while sharpening is obtained by shifting the center sample outward, if $X_{(t)} < X^* < X_{(N-t+1)}$.

Another equivalent definition is obtained by using lower and upper statistics defined as follows:

$$\begin{aligned} X^L &= \text{MED}\{X_{(s)}, X^*, X_{(t)}\}, \\ X^U &= \text{MED}\{X_{(N-t+1)}, X^*, X_{(N-s+1)}\}. \end{aligned}$$

Then the output of the LUM filter is given by

$$\text{LUM}(X_1, X_2, \ldots, X_N; s, t) = \begin{cases} X^L, & \text{if } X^* \leq (X^L + X^U)/2, \\ X^U, & \text{otherwise.} \end{cases} \qquad (3.12)$$

This definition is used in Algorithm 3.19.

Example 3.21. Let the 7-point input be $\mathbf{x} = (4, 2, 7, 9, 20, 5, 1)$ and let $s = 2$ and $t = 3$. The output of the

- LUM smoother is $\text{MED}\{X_{(s)}, X^*, X_{(N-s+1)}\} = \text{MED}\{2, 9, 9\} = 9$;

- LUM sharpener is $X^* = 9$, because $X^* > X_{(N-t+1)} = X_{(5)} = 7$;

- LUM filter is $X^* = 9$.

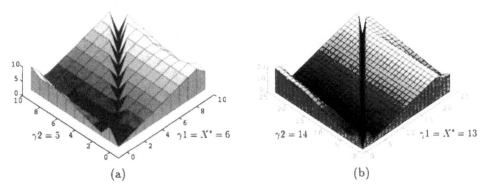

Figure 3.65. Functions $\mathcal{S}(\mathbf{x}^{r*})$ used in the filtering examples, (a) one-dimensional case, window length 11, (b) two-dimensional case, window size 5 × 5. The height of the mesh gives the output rank for the given feature vector.

3.10.2   Impulse and Step Response

It is easy to obtain permutation and RCRS filters that remove impulses. This is simply obtained by not using the extreme ranks 1 and N in the rank selection rule. The LUM smoother (LUM sharpener) removes impulses unless $s = 1$ $(t = 1)$. The LUM filter removes impulses if $s \neq 1$ (Exercise). It is also relatively easy to design permutation and RCRS filters that preserve step edges. Since the output of the permutation and RCRS filters is always one of the input samples, they cannot blur the step edges but they can reorder the samples almost in any order. The LUM smoothers, LUM sharpeners, and LUM filters all preserve step edges (Exercise).

3.10.3   Filtering Examples

For practical reasons we do not show examples of permutation filtering (it is not convenient to show in a book 11! let alone 25! permutations and the corresponding output ranks). By reducing the number of the permutations in some convenient way we would have ended up with some RCRS filter or some filter closely resembling it. For the RCRS filters the selected samples in the one-dimensional case were the center sample X^* and the sample before it, $\Gamma = \{5, 6\}$. The rank selection rule for the one-dimensional case is shown in Figure 3.65 (a). The selected samples in the two-dimensional case were the center sample X^* and the sample immediately to the right of it, $\Gamma = \{13, 14\}$. The rank selection rule for the two-dimensional signal is shown in Figure 3.65 (b). From Figure 3.65 one can see that for most of the feature vectors the output follows the rank of X^*, thus, the filters can preserve details well.

The smoothing properties of the LUM smoother are evident in the examples as well as the sharpening properties of the LUM sharpener (e.g., it is not able to remove noise). Furthermore, it is easily seen that the LUM filter combines both the LUM smoother and the LUM sharpener.

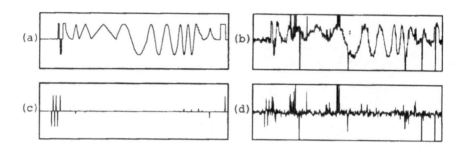

Figure 3.66. One-dimensional signals filtered by the RCRS filter of window length 11 and rank selection rule shown in Figure 3.65 (a).

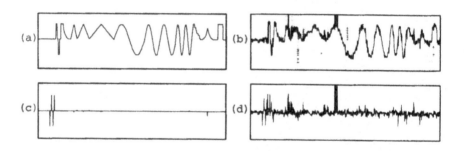

Figure 3.67. One-dimensional signals filtered by the LUM smoother of window length 11, $s = 3$.

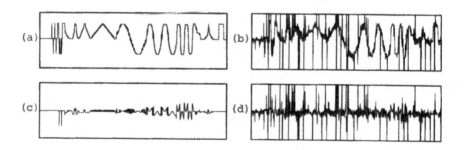

Figure 3.68. One-dimensional signals filtered by the LUM sharpener of window length 11, $t = 3$.

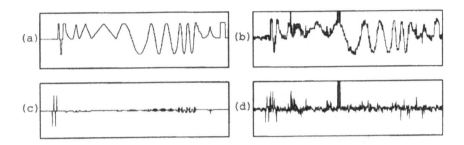

Figure 3.69. One-dimensional signals filtered by the LUM filter of window length 11, $s = 3$, $t = 5$.

Figure 3.70. "Geometrical" filtered by the RCRS filter of window size 5×5 and rank selection rule shown in Figure 3.65 (b).

Figure 3.71. "Geometrical" filtered by the LUM smoother of window size 5×5, $s = 6$.

Figure 3.72. "Geometrical" filtered by the LUM sharpener of window size 5×5, $t = 6$.

Figure 3.73. "Geometrical" filtered by the LUM filter of window size 5×5, $s = 6$, $t = 11$.

3.11 M-Filters

3.11.1 Principles and Properties

As indicated in Chapter 2, M-estimators are generalizations of maximum likelihood estimators. As we have emphasized, filtering can be understood as an estimation application. Therefore it becomes natural to exploit also M-estimators for filtering [63, 64]. Recall that a location M-estimator or an M-filter is defined by a function ρ on **R**:

$$M(X_1, X_2, \ldots, X_N; \rho) = \arg \min_{\theta \in \Theta} \sum_{i=1}^{N} \rho(X_i - \theta)$$

or by the partial derivative ψ (with respect to θ) of ρ: the estimate $M(X_1, X_2, \ldots, X_N)$ satisfies the equation

$$\sum_{i=1}^{N} \psi(X_i - \theta) = 0. \tag{3.13}$$

There exist several techniques of finding the output [40]. Here we derive one of them: By assuming $\lim_{x \to 0} \frac{\psi(x)}{x} = c$ we can write (3.13) in the form

$$\sum_{i=1}^{N} \psi(X_i - \theta) = \sum_{X_i \neq \theta} \frac{\psi(X_i - \theta)}{X_i - \theta}(X_i - \theta) = 0, \tag{3.14}$$

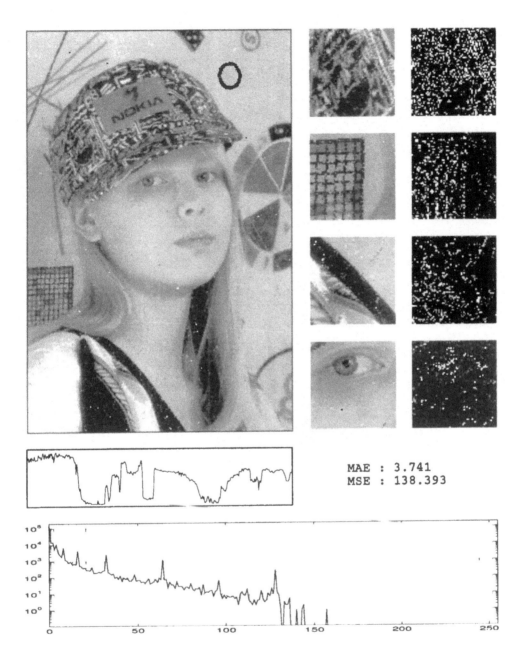

```
MAE : 3.741
MSE : 138.393
```

Figure 3.74. "Leena" filtered by the RCRS filter with window size 5 × 5 and rank selection rule shown in Figure 3.65 (b).

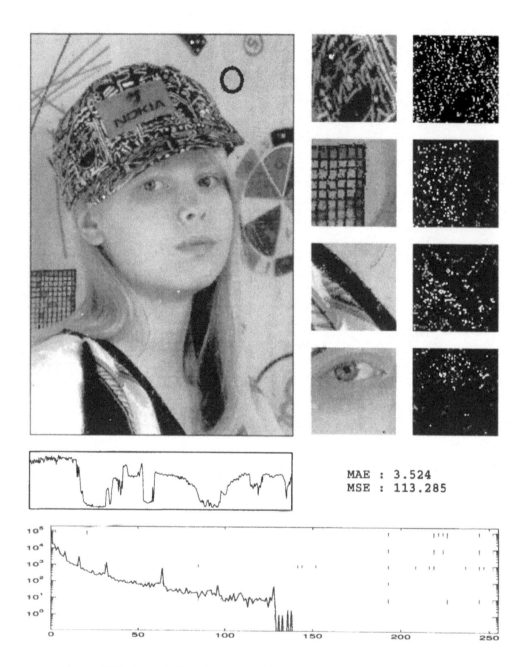

MAE : 3.524
MSE : 113.285

Figure 3.75. "Leena" filtered by the LUM smoother of window size 5×5, $s = 6$.

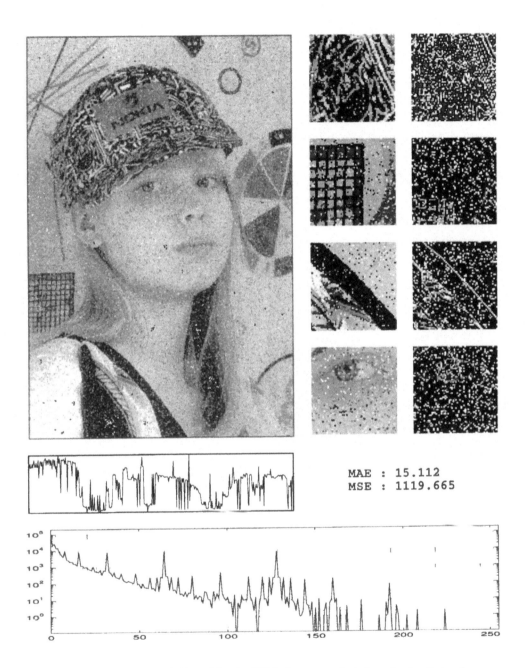

MAE : 15.112
MSE : 1119.665

Figure 3.76. "Leena" filtered by the LUM sharpener of window size 5×5, $t = 6$.

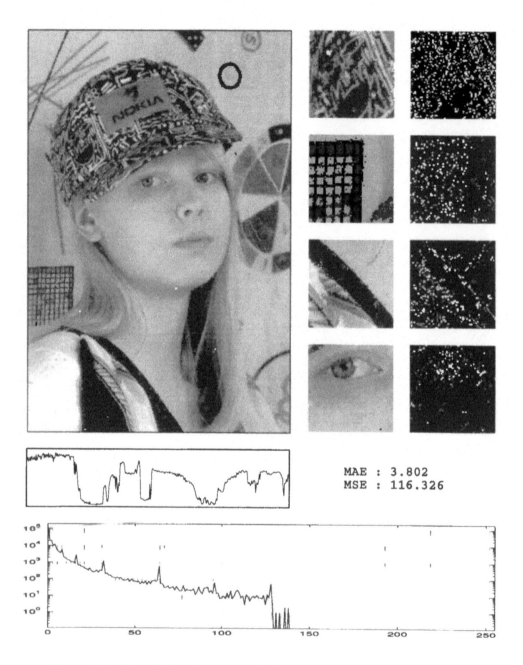

MAE : 3.802
MSE : 116.326

Figure 3.77. "Leena" filtered by the LUM filter of window size 5×5, $s = 6$, $t = 11$.

from which we can easily find an implicit formula for the estimate θ (Exercise):

$$\theta = \frac{\sum_{i=1}^{N} w_{i,\theta} X_i}{\sum_{i=1}^{N} w_{i,\theta}}, \tag{3.15}$$

where

$$w_{i,\theta} = \begin{cases} \frac{\psi(X_i - \theta)}{X_i - \theta}, & X_i \neq \theta, \\ c, & \text{otherwise.} \end{cases}$$

This can be used iteratively

$$\theta^{(k+1)} = \frac{\sum_{i=1}^{N} w_{i,\theta^{(k)}} X_i}{\sum_{i=1}^{N} w_{i,\theta^{(k)}}}.$$

A convergence proof is given in Reference [40]. The iteration limit $\theta^{(\infty)}$ is a solution of

$$\sum_{i=1}^{N} \psi(X_i - \theta) = 0.$$

This method is shown in Algorithm 3.20.

In Reference [64] the *M*-filters with

$$\psi(x) = \begin{cases} g(p), & x > p, \\ g(x), & |x| \leq p, \\ -g(p), & x < -p, \end{cases}$$

where $g(x)$ is a strictly increasing, odd, continuous function and p is some positive constant, are called *limiter type M-filters*. When $g(x) = ax$, a *standard type M-filter* is obtained. Limiter type *M*-filters have many desirable properties. The output obtained using (3.13) is unique and the filter can be robust. When p is chosen properly, it can reduce the magnitude of the impulses. The choice of p can be done by using some scale estimate obtained from *a priori* knowledge of the signal and noise distributions.

If we are willing to discard the magnitude of impulses entirely, *redescending M-filters* provide an alternative. For redescending *M*-filters their ψ-function vanishes outside some central region. Some popular redescending *M*-estimators/filters are listed below (see Figure 3.78):

- Skipped median, [30]

$$\psi_{\text{med}(r)}(x) = \begin{cases} \text{sign}(x), & 0 \leq |x| < r, \\ 0, & r \leq |x|. \end{cases}$$

- Hampel's three part redescending, [2]

$$\psi_{a,b,r}(x) = \begin{cases} x, & 0 \leq |x| < a, \\ a\,\text{sign}(x), & a \leq |x| < b, \\ a\frac{r-|x|}{r-b}\text{sign}(x), & b \leq |x| < r, \\ 0, & r \leq |x|. \end{cases}$$

- Andrew's sine, [2]

$$\psi_{\sin(a)}(x) = \begin{cases} \sin(x/a), & 0 \leq |x| < \pi a, \\ 0, & \pi a \leq |x|. \end{cases}$$

M-filter

Inputs: *NumberOfRows* × *NumberOfColumns* image
        Moving window $W, |W| = N = 2k + 1$
        Function $\psi(x)$
        Constant real number $c \neq 0$
        Constant natural number *NumberOfIterations*
Output: *NumberOfRows* × *NumberOfColumns* image

```
for i = 1 to NumberOfRows
    for j = 1 to NumberOfColumns
        place the window W at (i, j)
        store the image values inside W in x = (X₁, X₂, ..., X_N)
        find the median θ(0) of the values inside W
        for k = 1 to NumberOfIterations
            let Sum1 = 0,  Sum2 = 0
            for l = 1 to N
                if X_l ≠ θ(k − 1)
                    let Sum1 = Sum1 + X_l · (ψ(X_l − θ(k − 1)))/(X_l − θ(k − 1))
                    let Sum2 = Sum2 + (ψ(X_l − θ(k − 1)))/(X_l − θ(k − 1))
                else
                    let Sum1 = Sum1 + c · X_l
                    let Sum2 = Sum2 + c
            end
            let θ(k) = Sum1/Sum2
        end
        let Output(i, j) = θ(k)
    end
end
```

Algorithm 3.20. Algorithm for the *M*-filter.

- Tukey's biweight, [13]

$$\psi_{\text{bi}(r)}(x) = \begin{cases} x(r^2 - x^2)^2, & 0 \leq |x| < r, \\ 0, & r \leq |x|. \end{cases}$$

The corresponding filters are very robust but they have computational problems [40]. First of all, as the ψ function is not strictly monotone the solution of (3.13) might not be unique. This can be handled in several ways:

1. Find the global minimum of $\sum_{i=1}^{N} \rho(X_i - \theta)$;

2. Select the solution of $\sum_{i=1}^{N} \psi(X_i - \theta) = 0$ nearest to the sample median;

3. Use Newton's method, starting from the sample median;

4. Use the above iterative method, starting from the sample median;

5. Use some one-step method, starting from the sample median.

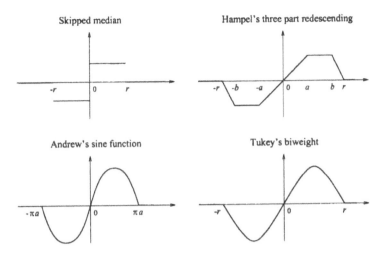

Figure 3.78. ψ-functions of some redescending M-filters.

The second problem is their sensitivity to an erroneous scale value. Some auxiliary scale estimate should be used to control the region of rejection.

Example 3.22.   Let the 5-point input be $\mathbf{x} = (2, 7, 9, 20, 5)$. Let us consider

- the standard type M-filter with $p = 5$ and $a = 1/5$;
- the Tukey's biweight M-filter with $r = 5$.

The plots of

$$\sum_{i=1}^{N} \psi(X_i - \theta) = \psi(2 - \theta) + \psi(7 - \theta) + \psi(9 - \theta) + \psi(20 - \theta) + \psi(5 - \theta)$$

are shown in Figure 3.79 below. The only zero for the standard type M-filter is at $\theta = 7$, which is the output, whereas we have various candidates for the output of the Tukey's biweight M-filter. Which one is found depends on the method used.

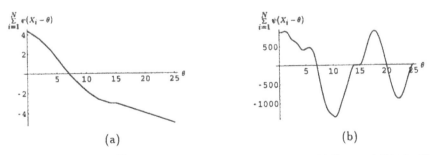

Figure 3.79. Plots of $\sum_{i=1}^{N} \psi(X_i - \theta)$ of (a) the standard type M-filter and (b) the Tukey's biweight M-filter. See Example 3.22 for the parameters used.

Gamma filters [8] are natural special cases of M-filters. We noticed in Chapter 2 that the mean is the θ minimizing

$$\sum_{i=1}^{N} |X_i - \theta|^2$$

and the median is the θ minimizing

$$\sum_{i=1}^{N} |X_i - \theta|.$$

What happens if we use powers other than 1 and 2? Naturally, new filters are obtained. Assume that $\gamma > 0$ and let $\hat{\theta}$ be the the value minimizing

$$\sum_{i=1}^{N} |X_i - \theta|^\gamma. \tag{3.16}$$

Then the following statements hold for $\hat{\theta}$: [8]:

1. $\hat{\theta}$ is the maximum likelihood estimate of the location parameter β on a random sample $\{X_1, X_2, \ldots, X_N\}$ from a population with the exponential density $f(x) = \alpha \exp\{-|x - \beta|^\gamma\}$, where α is the necessary scaling factor and $\gamma > 0$ (Exercise). The smaller the γ is, the stronger tails the density has.

2. If $\gamma > 1$, then $\hat{\theta}$ is unique (Exercise).

3. If $\gamma \leq 1$, then $\hat{\theta}$ is one of the samples $\{X_1, X_2, \ldots, X_N\}$.

4. If $\gamma \to 0$ then $\hat{\theta} \to X_k$ minimizing $\prod_{i \neq k} |X_i - X_k|$, i.e., $\hat{\theta}$ tends to some concentration of the samples when the density is very heavy-tailed.

5. If $\gamma \to \infty$ then $\hat{\theta} \to (X_{(1)} + X_{(N)})/2$, i.e., $\hat{\theta}$ tends to the midrange when the density is close to uniform.

The output of the γ filter is defined to be the sample $X_k \in \{X_1, X_2, \ldots, X_N\}$ minimizing (3.16). Thus, if $\gamma > 1$, the filter output may not be the maximum likelihood estimate of the location parameter but the sample value in the current window which is most likely the correct value of the location parameter (the ties must naturally be decided somehow). This deviation from the maximum likelihood principle simplifies significantly the implementation; see Algorithm 3.21. So, e.g., the gamma filter when $\gamma = 2$ is not the mean filter as the output must be one of the input samples, but it is shown that the output is the input sample which is closest to the mean [8].

It can be shown that the weighted mean (Exercise)

$$\theta = \frac{\sum_{i=1}^{N} w_i X_i}{\sum_{i=1}^{N} w_i}$$

minimizes

$$\sum_{i=1}^{N} w_i |X_i - \theta|^2,$$

and the weighted median minimizes

$$\sum_{i=1}^{N} w_i |X_i - \theta|.$$

Gamma filter

Inputs: *NumberOfRows* × *NumberOfColumns* image
        Moving window $W, |W| = N = 2k + 1$
        Real number γ
Output: *NumberOfRows* × *NumberOfColumns* image

```
for i = 1 to NumberOfRows
    for j = 1 to NumberOfColumns
        place the window W at (i, j)
        store the image values inside W in x = (X₁, X₂, ..., Xₙ)
        let SmallestSum = LargestPossibleValue
        for k = 1 to N
            let Sumₖ = Σᴺᵢ₌₁ |Xᵢ − Xₖ|ᵞ
            if Sumₖ < SmallestSum
                let SmallestSum = Sumₖ
                let OutputCandidate = Xₖ
            end
        end
        let Output(i, j) = OutputCandidate
    end
end
```

Algorithm 3.21. Algorithm for the Gamma filter.

In an analogous manner, the weighted gamma filter output is defined to be the sample $X_k \in \{X_1, X_2, \ldots, X_N\}$ minimizing

$$\sum_{i=1}^{N} w_i |X_i - \theta|^\gamma.$$

By weights different emphasis can be assigned to different samples. Again, the idea is to gain an improved detail preservation by using these weights.

This can be furthermore generalized by changing the distance function $|a - b|^\gamma$ into a more general distance function $g(a, b)$ satisfying

- $g(a, a) = 0$ for all a;

- $g(a, b) \geq 0$ for all a, b;

- $g(a, b) \geq g(c, d)$ if $|a - b| \geq |c - d|$ for all a, b, c, d;

- $g(a, b) = g(b, a)$ for all a, b.

Thus, e.g., for 8-bit images with 256 gray-level values the possible distances between two samples are within the range $0 \ldots 255$ and the number of different pairs a, b for which the distance function must be defined is $256 \cdot 255$. The most natural simplification is obtained by using a function $g(a, b) = g(|a - b|)$ satisfying

- $g(0) = 0$;

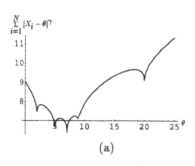
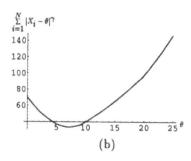

(a)   (b)

Figure 3.80. Plots of $\sum_{i=1}^{N} |X_i - \theta|^\gamma$ (a) $\gamma = 0.3$ and (b) $\gamma = 1.2$. See Example 3.23.

- $g(x)$ is increasing.

Thus, for 8-bit images one has to define values $g(0), g(1), \ldots, g(255)$ satisfying the above two conditions. These distance functions can be understood as forming a very general class of M-filters suited for signal processing purposes.

Example 3.23. Let the 5-point input be $\mathbf{x} = (2, 7, 9, 20, 5)$. Let us consider gamma values 0.3 and 1.2. The plots of

$$\sum_{i=1}^{N} |X_i - \theta|^\gamma = |2 - \theta|^\gamma + |7 - \theta|^\gamma + |9 - \theta|^\gamma + |20 - \theta|^\gamma + |5 - \theta|^\gamma$$

are shown in Figure 3.80.

For $\gamma = 0.3$ the function obtains its minimum at 7, one of the samples, and for $\gamma = 1.2$ the nearest sample to the minimum is 7.

3.11.2   Impulse and Step Response

Because of the wide variety of ψ-functions, it is impossible to categorize M-filters into impulse removing/not-removing types. The same holds for the step response. Typically, the redescending and standard type M-filters with properly chosen parameters can remove impulses and preserve step edges reasonably well.

Consider gamma-filtering and let all weights be equal to 1. In the case of an impulse, we have two values inside the window, 0 and a, and these are the only values to be considered as output candidates. For 0 the sum in (3.16) attains value $(N-1)|0-0|^\gamma + |a-0|^\gamma = |a|^\gamma$ and for a the sum in (3.16) attains value $(N-1)|0-a|^\gamma + |a-a|^\gamma = (N-1)|a|^\gamma$, where the case $\theta = 0$ is smaller if $N > 1$. Thus, the impulse is removed. Consider then the step response. Let α samples inside the window have the value a and the rest $N - \alpha$ samples the value b. The sums (3.16) for the possible outputs a and b equal $L(a) = (N-\alpha)|b-a|^\gamma$ and $L(b) = \alpha|b-a|^\gamma$, respectively. The output is a, if $N - \alpha < \alpha$, i.e., $\alpha < N/2$, which is true before the edge, and after the edge the output is b. Thus, the step edge is preserved.

One can also easily see that if one sample is weighted heavily (its weight is greater than the sum of the other weights), the impulse may stay. A sufficient condition for the step preservation is the use of symmetrical weights: $a_i = a_{N-i+1}$ for all i.

3.11.3 Filtering Examples

We show the results using the standard type M-filter with $p = 10$ and $a = 1/10$ and the Tukey's biweight M-filter with $r = 5$. The computational problem of the Tukey's biweight M-filter has been solved in these examples by using the following one-step version:

$$M(X_1, X_2, \ldots, X_N) = med + 1.483\text{MAD} \sum_{i=1}^{N} \psi \left(\frac{X_i - med}{1.483\text{MAD}} \right) \bigg/ \sum_{i=1}^{N} \psi' \left(\frac{X_i - med}{1.483\text{MAD}} \right).$$

For gamma filters we show the results using two different values of γ: 0.7 and 1.2.

The results of all these filters are relatively close to the results of the median filter. From the results of the Tukey's biweight we can also see a strong low-pass effect. One may understand that the Tukey's biweight M-filter acts like the mean filter for high frequencies. The results are sharpest for the gamma filter with $\gamma = 0.7$.

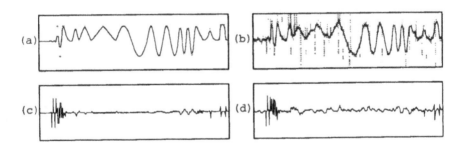

Figure 3.81. One-dimensional signals filtered by the standard type M-filter of window length 11, $p = 10$ and $a = 1/10$.

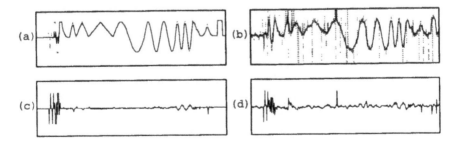

Figure 3.82. One-dimensional signals filtered by the Tukey's biweight M-filter of window length 11 and $r = 5$.

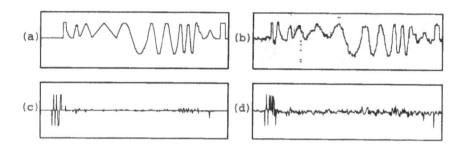

Figure 3.83. One-dimensional signals filtered by the gamma filter of window length 11, $\gamma = 0.7$.

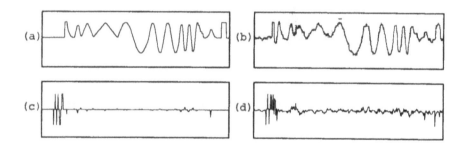

Figure 3.84. One-dimensional signals filtered by the gamma filter of window length 11, $\gamma = 1.2$.

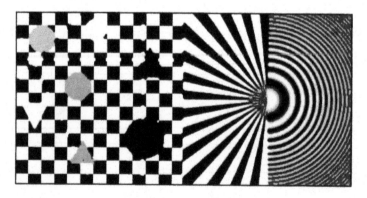

Figure 3.85. "Geometrical" filtered by the standard type M-filter of window size 5×5, $p = 10$ and $a = 1/10$.

Figure 3.86. "Geometrical" filtered by the Tukey's biweight M-filter window size 5×5 and $r = 5$.

Figure 3.87. "Geometrical" filtered by the gamma filter of window size 5×5, $\gamma = 0.7$.

Figure 3.88. "Geometrical" filtered by the gamma filter of window size 5×5, $\gamma = 1.2$.

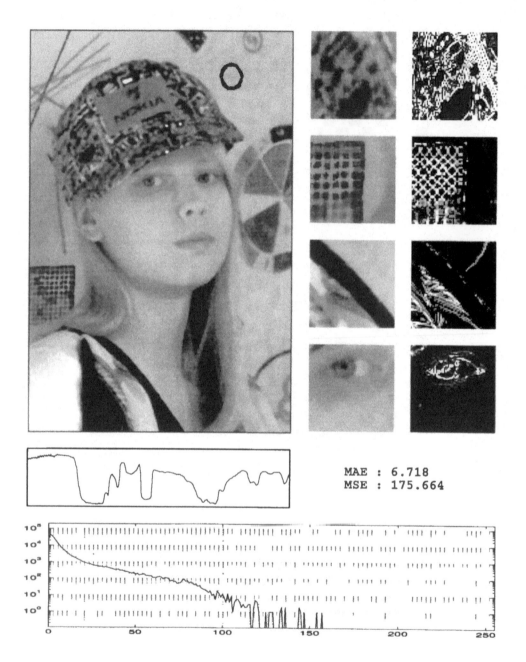

MAE : 6.718
MSE : 175.664

Figure 3.89. "Leena" filtered by the standard type M-filter of window size 5×5, $p = 10$ and $a = 1/10$.

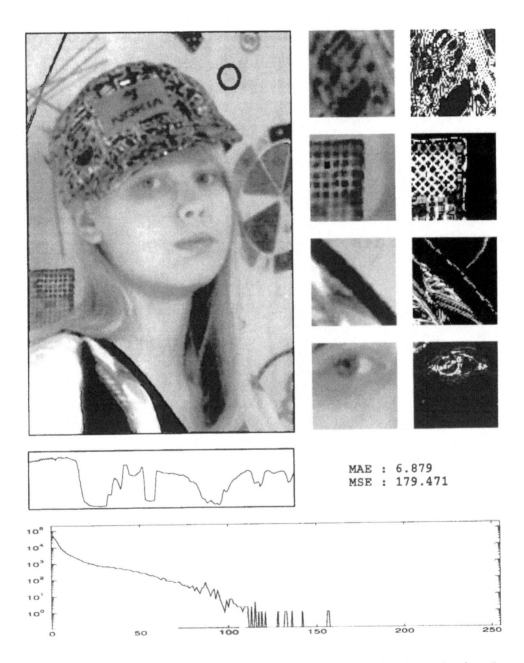

MAE : 6.879
MSE : 179.471

Figure 3.90. "Leena" filtered by the Tukey's biweight M-filter of window size 5×5 and $r = 5$.

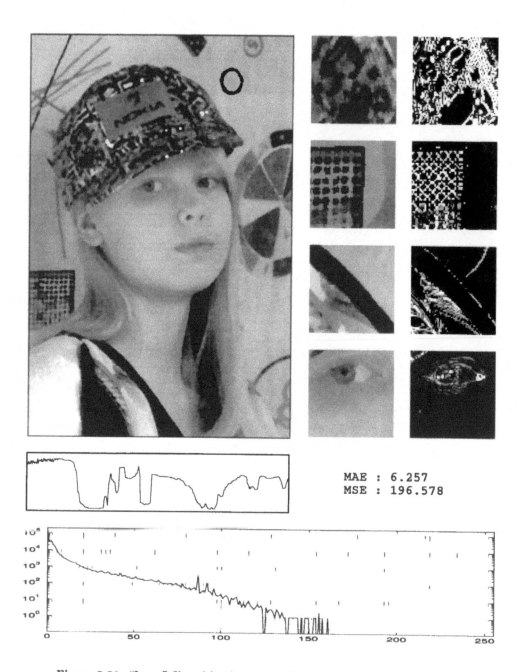

MAE : 6.257
MSE : 196.578

Figure 3.91. "Leena" filtered by the gamma filter of window size 5×5, $\gamma = 0.7$.

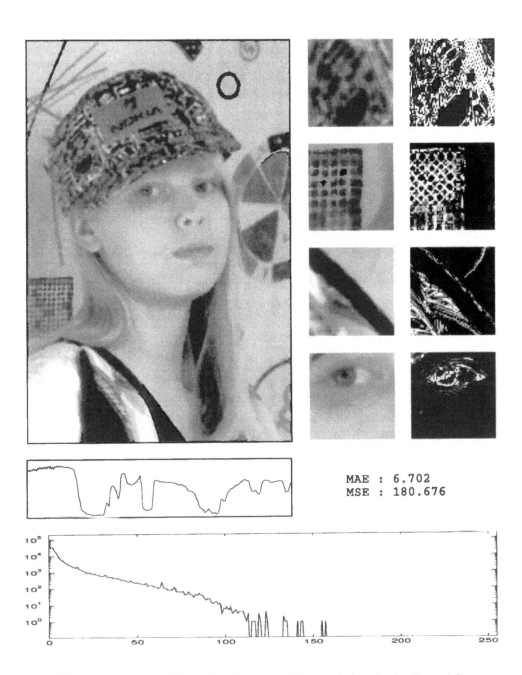

MAE : 6.702
MSE : 180.676

Figure 3.92. "Leena" filtered by the gamma filter of window size 5×5, $\gamma = 1.2$.

Wilcoxon filter

Inputs: *NumberOfRows* × *NumberOfColumns* image
       Moving window $W, |W| = N = 2k + 1$
Output: *NumberOfRows* × *NumberOfColumns* image

```
for i = 1 to NumberOfRows
    for j = 1 to NumberOfColumns
        place the window W at (i, j)
        store the image values inside W in x = (X₁, X₂, ..., Xₙ)
        let m = 1
        for s = 1 to N
            for t = s to N
                let Ave(m) = (Xₛ + Xₜ)/2
                let m = m + 1
            end
        end
        find the median Med of the vector (Ave(1), Ave(2), ..., Ave(m − 1))
        let Output(i, j) = Med
    end
end
```

Algorithm 3.22. Algorithm for the Wilcoxon filter.

3.12 *R*-Filters

3.12.1 Principles and Properties

R-filters are based on the well-known *R*-estimators, which in turn are based on rank tests [37]. *R*-estimators are known to be robust estimators, and this has attracted the signal processing society to apply them also to filtering, as *M*- and *L*-estimators have been applied.

The first introduced *R*-filter is the *Wilcoxon filter* [23]:

$$\text{Wil}(X_1, X_2, \ldots, X_N) \;=\; \text{MED}\left\{\frac{X_i + X_j}{2} : 1 \le i \le j \le N\right\} \tag{3.17}$$

$$=\; \text{MED}\left\{\frac{X_{(i)} + X_{(j)}}{2} : 1 \le i \le j \le N\right\}. \tag{3.18}$$

Thus, the Wilcoxon filter consists of linear operations, called *Walsh averages* $W_{i,j} = (X_i + X_j)/2$ or *ordered Walsh averages* $W_{(i,j)} = (X_{(i)} + X_{(j)})/2$, and of a nonlinear operation, the median. The Wilcoxon filter is in fact a customary version of the well-known *Hodges-Lehmann* estimator [37] applied to filtering. In the Hodges-Lehmann estimator all the N^2 Walsh averages are used without requiring $1 \le i \le j \le N$. The name Wilcoxon comes from the fact that the Hodges-Lehmann estimator is derived from the Wilcoxon test.

It is easy to see that the Wilcoxon filter actually is a special case of the median hybrid filters. The difference is that typically in a median hybrid filter the subwindows are connected, which is not true now.

Example 3.24. Let the 5-point input be $\mathbf{x} = (2, 7, 9, 20, 5)$. Then the (ordered) Walsh averages are $(2+2)/2, (2+7)/2, (2+9)/2, (2+20)/2, (2+5)/2, (7+7)/2, (7+9)/2, (7+20)/2, (7+5)/2, (9+9)/2, (9+20)/2, (9+5)/2, (20+20)/2, (20+5)/2, (5+5)/2$. Their median 7 is the output of the Wilcoxon filter.

One main disadvantage of the Wilcoxon filter is its computational complexity. It is easy to see that the Wilcoxon filter computes the median of $N(N+1)/2$ Walsh averages, which is a significant increase in number compared with the median filter. However, the computational complexity can be reduced [23, 59]. For example, consider the case $N = 5$. The Wilcoxon filter output is the median of the $N(N+1)/2 = 15$ ordered Walsh averages shown below:

	$X_{(1)}$	$X_{(2)}$	$X_{(3)}$	$X_{(4)}$	$X_{(5)}$
$X_{(1)}$	$\frac{X_{(1)}+X_{(1)}}{2}$	$\frac{X_{(1)}+X_{(2)}}{2}$	$\frac{X_{(1)}+X_{(3)}}{2}$	$\frac{X_{(1)}+X_{(4)}}{2}$	$\frac{X_{(1)}+X_{(5)}}{2}$
$X_{(2)}$		$\frac{X_{(2)}+X_{(2)}}{2}$	$\frac{X_{(2)}+X_{(3)}}{2}$	$\frac{X_{(2)}+X_{(4)}}{2}$	$\frac{X_{(2)}+X_{(5)}}{2}$
$X_{(3)}$			$\frac{X_{(3)}+X_{(3)}}{2}$	$\frac{X_{(3)}+X_{(4)}}{2}$	$\frac{X_{(3)}+X_{(5)}}{2}$
$X_{(4)}$				$\frac{X_{(4)}+X_{(4)}}{2}$	$\frac{X_{(4)}+X_{(5)}}{2}$
$X_{(5)}$					$\frac{X_{(5)}+X_{(5)}}{2}$

From this table it is clear that $W_{(1,1)}, W_{(1,2)}, W_{(1,3)}, W_{(1,4)}, W_{(2,2)}$, and $W_{(2,3)}$ are smaller than at least 8 of the other ordered Walsh averages, so that they are all smaller than (or equal to) the median. Similarly, $W_{(2,5)}, W_{(3,4)}, W_{(3,5)}, W_{(4,4)}, W_{(4,5)}$, and $W_{(5,5)}$ are larger than (or equal to) the median. Thus, we can exclude an equal number of values from below and above the median value. The output is the median of the remaining ordered Walsh averages, i.e., the Wilcoxon filter is given by

$$\text{Wil}(X_1, X_2, \ldots, X_5) = \text{MED}\left\{X_{(3)}, \frac{X_{(1)} + X_{(5)}}{2}, \frac{X_{(2)} + X_{(4)}}{2}\right\}.$$

Thus, the calculation of the median of 15 ordered Walsh averages is reduced to the calculation of the median of 3 ordered Walsh averages.

The Wilcoxon filter can also be Winsorized or trimmed [25, 26]. The first form of Winsorizing implies Walsh averages being computed only for samples whose ranks are within a fixed maximum distance from each other:

$$\text{RW-Wil}(X_1, X_2, \ldots, X_N; r) = \text{MED}\left\{\frac{X_{(i)} + X_{(j)}}{2} : j - i < r, 1 \le i \le j \le N\right\}.$$

By letting r range from 1 to N, the *Rank Winsorized Wilcoxon filter* ranges from the median filter to the Wilcoxon filter.

Example 3.25. Let the 5-point input be $\mathbf{x} = (2, 7, 9, 20, 5)$. The ordered vector is $(2, 5, 7, 9, 20)$. If $r = 2$, the utilized Walsh averages are $W_{(1,1)} = (2+2)/2, W_{(1,2)} = (2+5)/2, W_{(2,2)} = (5+5)/2, W_{(2,3)} = (5+7)/2, W_{(3,3)} = (7+7)/2, W_{(3,4)} = (7+9)/2, W_{(4,4)} = (9+9)/2, W_{(4,5)} = (9+20)/2, W_{(5,5)} = (20+20)/2$. Their median $W_{(3,3)} = 7$ is the output of the Rank Winsorized Wilcoxon filter.

Another way of trimming is to use Walsh averages computed only for samples being within a fixed maximum temporal distance of each other, e.g., in the one-dimensional case

$$\text{TW-Wil}(X_1, X_2, \ldots, X_N; t) = \text{MED}\left\{\frac{X_i + X_j}{2} : j - i < t, 1 \le i \le j \le N\right\}.$$

This *Time Winsorized Wilcoxon filter* ranges from the median filter to the Wilcoxon filter when t ranges from 1 to N.

Example 3.26. Let the 5-point input be x = $(2, 7, 9, 20, 5)$. If $t = 2$, the utilized Walsh averages are $W_{1,1} = (2+2)/2, W_{1,2} = (2+7)/2, W_{2,2} = (7+7)/2, W_{2,3} = (7+9)/2, W_{3,3} = (9+9)/2, W_{3,4} = (9+20)/2, W_{4,4} = (20+20)/2, W_{4,5} = (20+5)/2, W_{5,5} = (5+5)/2$. Their median $W_{2,3} = 8$ is the output of the Time Winsorized Wilcoxon filter.

Naturally, we can employ both time and rank Winsorizing at the same time. Another generalization of the Wilcoxon filter can be obtained by using some more general m-wise averages instead of pairwise averages [26]. Still another way to use Winsorizing/trimming is to combine the ideas of the modified trimmed mean filtering and the Wilcoxon filtering: find the median of the window first and use the samples lying within a fixed distance from it for Wilcoxon filtering [23].

A filter related to the Wilcoxon filter is the *Hodges-Lehmann D-filter* [54]:

$$\text{H-L}(X_1, X_2, \ldots, X_N) = \text{MED}\left\{\frac{X_{(i)} + X_{(N-i+1)}}{2} : 1 \le i \le k+1\right\}.$$

Thus, the Hodges-Lehmann D-filter is a special case of the Wilcoxon filter and is the median of the quasi-midranges $(X_{(i)} + X_{(N-i+1)})/2$. In fact, we now see that the Hodges-Lehmann D-filter coincides with the Wilcoxon filter if $N = 5$ (compare the definition of the first one with the reduced formula for the latter one). The Hodges-Lehmann D-filter is known in statistics as the *Bickel-Hodges estimator*. In Reference [54] a double window technique has also been introduced for the Hodges-Lehmann D-filter. This technique is in many ways similar to the double window technique discussed in Section 3.2.

Example 3.27. Let the 5-point input be x = $(2, 7, 9, 20, 5)$. The output of the Hodges-Lehmann D-filter is now MED$\{(2+20)/2, (5+9)/2, (7+7)/2\}$ = MED$\{11, 7, 7\}$ = 7.

3.12.2 Impulse and Step Response

The impulsive value can be at most in N of the Walsh averages. Since $N < (N(N+1)/2)/2$ if $N > 3$, this means for $N > 3$ that the impulse response of the Wilcoxon filter is a zero-valued sequence. As we pointed out, when r ranges from 1 to N, the Rank Winsorized Wilcoxon filter ranges from the median filter to the Wilcoxon filter; when t ranges from 1 to N the Time Winsorized Wilcoxon filter ranges from the median filter to the Wilcoxon filter. Thus, the Winsorized Wilcoxon filters offer a compromise in performance between the Wilcoxon and the median filter. Thus, intuitively the Winsorized Wilcoxon filters can remove impulses completely. The verification is left as an exercise.

Wilcoxon filters generally smear step edges. This is due to the averaging of all possible pairs before the median operation. However, the Wilcoxon filter does not smear edges as badly as the mean filter does. The fact that the Winsorized Wilcoxon filters offer a compromise in performance between the Wilcoxon and the median filter can be seen from their step responses. The closer we are to the median, the better the steps are preserved.

Hodges-Lehmann D-filter

Inputs: *NumberOfRows* × *NumberOfColumns* image
            Moving window $W, |W| = N = 2k + 1$
Output: *NumberOfRows* × *NumberOfColumns* image

```
for i = 1 to NumberOfRows
    for j = 1 to NumberOfColumns
        place the window W at (i, j)
        store the image values inside W in x = (X₁, X₂, ..., Xₙ)
        sort x, store the result in y = (X₍₁₎, X₍₂₎, ..., X₍ₙ₎)
        for m = 1 to k
            let Ave(m) = (X₍ₘ₎ + X₍ₙ₋ₘ₊₁₎)/2
        end
        find the median Med of the vector (Ave(1), Ave(2), ..., Ave(k), X₍ₖ₊₁₎)
        let Output(i, j) = Med
    end
end
```

Algorithm 3.23. Algorithm for the Hodges-Lehmann D-filter.

3.12.3   Filtering Examples

From the examples it is easy to see how both the Wilcoxon filter and the Hodges-Lehmann D-filter smear, or change into two-step versions, step edges. It is also seen that the Hodges-Lehmann D-filter is slightly worse in noise removal than the Wilcoxon filter, whereas the overall quality of the filtered signals is very similar.

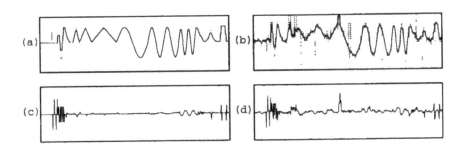

Figure 3.93. One-dimensional signals filtered by the Wilcoxon filter of window length 11.

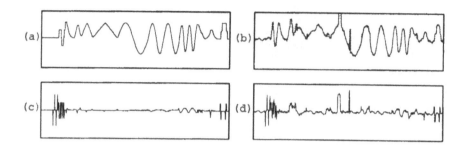

Figure 3.94. One-dimensional signals filtered by the Hodges-Lehmann D-filter of window length 11.

Figure 3.95. "Geometrical" filtered by the Wilcoxon filter of window size 5×5.

Figure 3.96. "Geometrical" filtered by the Hodges-Lehmann D-filter of window size 5×5.

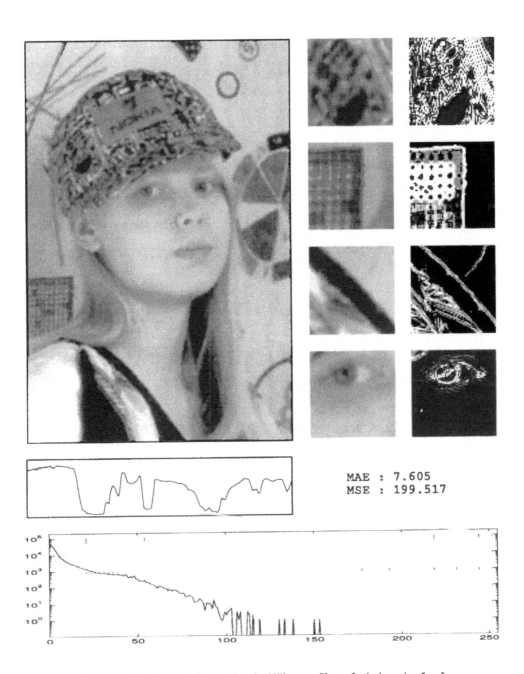

MAE : 7.605
MSE : 199.517

Figure 3.97. "Leena" filtered by the Wilcoxon filter of window size 5×5.

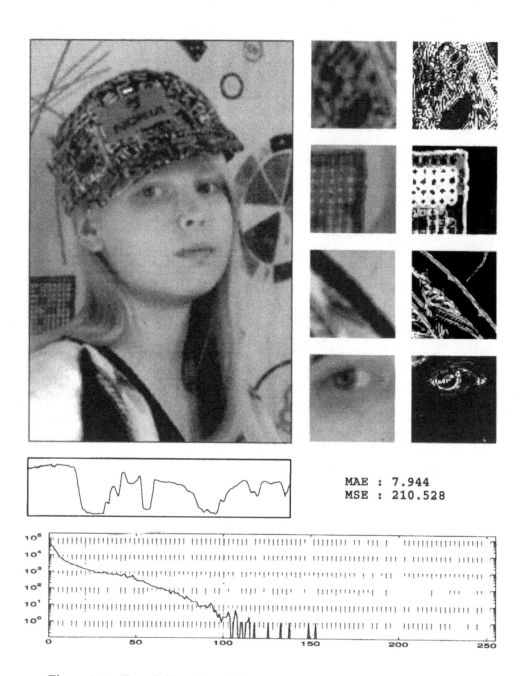

MAE : 7.944
MSE : 210.528

Figure 3.98. "Leena" filtered by the Hodges-Lehmann D-filter of window size 5×5.

3.13   Weighted Majority with Minimum Range Filters

3.13.1   Principles and Properties

The *Weighted Majority of m values with Minimum Range (WMMR^m) filters* have been developed in order to restore a wide class of distorted edges, while smoothing noisy nonedge regions and rejecting impulses [72]. A WMMR^m filter will select m $(k < m \leq 2k + 1)$ of the sorted values inside the moving window with the smallest range, and weight these values. Thus, first the index J minimizing $X_{(J+m-1)} - X_{(J)}$ is found and then the weighted mean of the samples $X_{(J)}, X_{(J+1)}, \ldots, X_{(J+m-1)}$ with weights $a_1, a_2, \ldots, a_m$ is calculated and gives the output. Clearly, J need not be unique. In that case, the output is given by the mean of the weighted means corresponding to all the values minimizing the range. The procedure is illustrated below for the case of $N = 7, m = 4$:

i	1	2	3	4	5	6	7
X_i	2	0	9	5	7	3	7
$X_{(i)}$	0	2	3	5	7	7	9
$X_{(i+3)} - X_{(i)}$	5	5	**4**	**4**			

Thus, both indices 3 and 4 minimize the range. If the weights are given by $(0, 0.5, 0.25, 0.25)$, then the output is given by

$$\frac{(0 \cdot 3 + 0.5 \cdot 5 + 0.25 \cdot 7 + 0.25 \cdot 7) + (0 \cdot 5 + 0.5 \cdot 7 + 0.25 \cdot 7 + 0.25 \cdot 9)}{2} = 6.75.$$

Hence, the WMMR^m is actually based on the trimming ideology, the major difference between it and the α-trimmed mean and modified trimmed mean filters being in the decision of which samples will be trimmed out. Here, the samples are trimmed out which do not belong to the best concentration of the data. An algorithm for the WMMR^m filter is given in Algorithm 3.24.

If the sum of the weights is unity the filters will be called *normalized Weighted Majority of m values with Minimum Range filters*. The *shorth* [105] is a special case of weighted majority with minimum range filters. The shorth is defined as the mean of the shortest half of the samples, i.e., it is the WMMR^m filter for $m = k + 1$ and constant weights $1/(k+1)$. Another special case is the well-known estimator *least median of squares* which is shown to be the midpoint of the shortest half of the samples [29, 93]. Thus, it is the WMMR^m filter for $m = k + 1$ and weights $1/2, 0, 0, \ldots, 0, 1/2$.

3.13.2   Impulse and Step Response

The impulse will never be in the smallest range unless $m = N$, in which case the WMMR^m filter is the weighted mean filter. Thus, the impulse will be removed by the WMMR^m filters. It is easy to show that if $m = k + 1$, the step edges are preserved using the normalized WMMR^m filter since the minimum range lies on the proper side of the edge. The $m > k + 1$ case is left as an exercise.

Weighted Majority of m values with Minimum Range filter

Inputs: *NumberOfRows* × *NumberOfColumns* image
         Moving window W, $|W| = N = 2k + 1$
         Natural number m
         Weight vector $\mathbf{a} = (a_1, a_2, \ldots, a_m)$
Output: *NumberOfRows* × *NumberOfColumns* image

```
for i = 1 to NumberOfRows
    for j = 1 to NumberOfColumns
        place the window W at (i, j)
        store the image values inside W in x = (X₁, X₂, ..., X_N)
        sort x, store the result in y = (X_(1), X_(2), ..., X_(N))
        let FoundMin = X_(N) − X_(1)
        for J = 1 to N − m + 1
            let Dist = X_(J+m−1) − X_(J)
            if Dist < FoundMin
                let FoundJs(1) = J
                let NumberOfFounds = 1
                let FoundMin = Dist
            else if Dist == FoundMin
                let NumberOfFounds = NumberOfFounds + 1
                let FoundJs(NumberOfFounds) = J
            end
        end
        let Sum = 0
        for k = 1 to NumberOfFounds
            for l = 1 to m
                let Sum = Sum + a_l · X_(FoundJs(k)+l−1)
            end
        end
        let Output(i, j) = Sum/NumberOfFounds
    end
end
```

Algorithm 3.24. Algorithm for the Weighted Majority of m values with Minimum Range filter.

3.13.3 Filtering Examples

We show examples using normalized WMMR^m filters where $m = k + 1$ and the weights are $a_i = 1/m$ for all i. Thus, these filters are also shorth filters.

The edge-enhancing property is clearly seen in the figures as well as the relatively strong blurring.

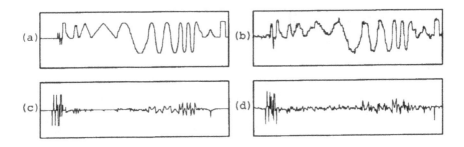

Figure 3.99. One-dimensional signals filtered by the WMMR^m filters of window length 11, $m = 6$, $a_i = 1/6$ for all i.

Figure 3.100. "Geometrical" filtered by the WMMR^m filters of window size 5×5, $m = 13$, $a_i = 1/13$ for all i.

3.14 Nonlinear Mean Filters

3.14.1 Principles and Properties

The arithmetic mean is not the only well-known mean. Other widely used means include, e.g., harmonic, geometric, and L_p means. Thus, an obvious attempt to solve filtering problems is to apply different means to signals in the same way as we apply the standard arithmetic mean. Some generalized mean filters have been introduced in Reference [53]. In Reference [88] a general form of nonlinear mean filters has been introduced. This form contains the filters in [53] as special cases. The nonlinear mean of the numbers $X_1, X_2, \ldots, X_\cdot$ is given by

$$y = g^{-1}\left(\frac{\sum_{i=1}^N a_i g(X_i)}{\sum_{i=1}^N a_i}\right),$$

where $g(x)$ is a function $g : \mathbf{R} \to \mathbf{R}$ and the a_is are weights.

An intuitive interpretation of this operation is shown in Figure 3.102. First, the signal values are transformed by $g(\cdot)$ into new values. Then, the weighted average of the new values is computed. Finally, the output is obtained by applying the inverse transformation to the weighted average. Naturally, after the transformation $g(\cdot)$ some other operation instead of the averaging can be used, e.g., some nonlinear filter.

Consider first the case where $a_i = a_j$ for all i, j. The following choices of $g(x)$ produce

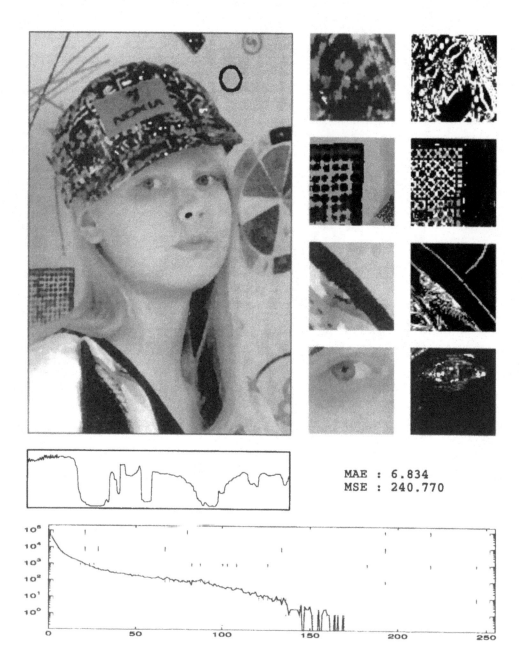

MAE : 6.834
MSE : 240.770

Figure 3.101. "Leena" filtered by the WMMR^m with window size 5×5, $m = 13$, $a_i = 1/13$ for all i.

Nonlinear mean filter

Inputs: *NumberOfRows* × *NumberOfColumns* image
Moving window $W, |W| = N = 2k + 1$
Functions $g(x), g^{-1}(x)$
Weight vector $a = (a_1, a_2, \ldots, a_N)$
Output: *NumberOfRows* × *NumberOfColumns* image

let $Sum2 = \sum_{i=1}^{N} a_i$
for $i = 1$ to *NumberOfRows*
    for $j = 1$ to *NumberOfColumns*
        $Sum1 = 0$
        place the window W at (i, j)
        for every value X_i of the image inside W
            let $Sum1 = Sum1 + a_i g(X_i)$
        end
        let $Output(i, j) = g^{-1}(Sum1/Sum2)$
    end
end

Algorithm 3.25. Algorithm for the nonlinear mean filter.

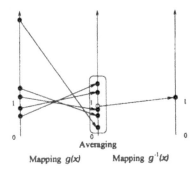

Mapping *g(x)*   Averaging   Mapping $g^{-1}(x)$

Figure 3.102. Nonlinear mean filtering.

interesting filters:

$$g(x) = \begin{cases} x, & \text{the } \textit{(arithmetic) mean} \text{ filter MEAN,} \\ 1/x, & \text{the } \textit{harmonic mean} \text{ filter HARM,} \\ \ln x, & \text{the } \textit{geometric mean} \text{ filter GEOM,} \\ x^p, p \in \mathbf{R} \setminus \{-1, 0, 1\}, & \text{the } L_p \textit{ mean} \text{ filter.} \end{cases}$$

Note that the base in the geometric mean filter does not need to be e. See Algorithm 3.25 for the algorithm of a nonlinear mean filter defined by the function $g(\cdot)$. The function $g(x)$ and its inverse can be given, e.g., in the form of a look-at table.

If we use signal dependent weights, $a_i = X_i^p, i = 1, 2, \ldots, N$, and the function $g(x) = x$, we obtain the *contraharmonic filter* CONTR$_p$.

These nonlinear means have the following well-known order property for $p > 0$ [47]. Let MIN and MAX denote the minimum and maximum values of the samples. Then

$$\text{MIN} \le \text{CONTR}_{-p} \le L_{-p} \le \text{HARM} \le \text{GEOM} \le \text{MEAN} \le L_p \le \text{CONTR}_p \le \text{MAX.}$$
(3.19)

Thus, the nonlinear mean filters $\text{CONTR}_{-p}, L_{-p}, \text{HARM}$, and GEOM tend to reduce the average signal level, while the nonlinear mean filters CONTR_p and L_p tend to increase the signal level.

Example 3.28. Let the 5-point input be $\mathbf{x} = (2, 7, 9, 20, 5)$. The output of the

- harmonic mean filter is $5/(1/2 + 1/7 + 1/9 + 1/20 + 1/5) \approx 4.98$;

- geometric mean filter is $\exp\{(\ln 2 + \ln 7 + \ln 9 + \ln 20 + \ln 5)/5\} \approx 6.61$;

- L_2 mean is $\sqrt{(2^2 + 7^2 + 9^2 + 20^2 + 5^2)/5} \approx 10.57$;

- L_{-2} mean is $\sqrt{5/(2^{-2} + 7^{-2} + 9^{-2} + 20^{-2} + 5^{-2})} \approx 3.92$;

- contraharmonic mean ($p = 2$) is $(2^3 + 7^3 + 9^3 + 20^3 + 5^3)/(2^2 + 7^2 + 9^2 + 20^2 + 5^2) \approx 16.47$;

- contraharmonic mean ($p = -2$) is $(2^{-1} + 7^{-1} + 9^{-1} + 20^{-1} + 5^{-1})/(2^{-2} + 7^{-2} + 9^{-2} + 20^{-2} + 5^{-2}) \approx 3.09$.

The geometric mean has a very important relationship with *homomorphic filters*. Homomorphic filters are used in cases where the noise is multiplicative [83]. A typical example is source illumination $E(x, y)$, which ,with object reflectance $r(x, y)$, contributes to image formation in a multiplicative way. The observed image is

$$i(x, y) = r(x, y)E(x, y)$$

and we are willing to extract $r(x, y)$ from it. When a logarithmic nonlinearity is applied to the image, the multiplicative "noise" is transformed into additive noise

$$\ln i(x, y) = \ln r(x, y) + \ln E(x, y).$$

This noise $\ln E(x, y)$ is then attenuated by a filter which gives the signal

$$t(x, y) \approx \ln r(x, y)$$

and the inverse transform (exp) is used to obtain the desired image

$$\hat{r}(x, y) = \exp\{t(x, y)\} \approx r(x, y).$$

3.14.2   Impulse and Step Response

Filtering by nonlinear means MEAN, HARM, GEOM, L_p, and CONTR_p tends to reduce the magnitude of impulses, which can be seen from (3.19). However, they tend to spread impulses slightly into the surrounding samples. Furthermore, the major drawback of nonlinear mean filtering for impulse removal is that they treat positive and negative impulses in a different way. Moreover, for the other-sided impulses, their magnitude also is crucial for the result. This can be seen from the following simple example: Consider geometric mean filtering and use logarithm to the base 10. Assume that out of the 11 inputs 10 have the value 100 and the 11th input has the value $a \in [0, 255]$. By definition the output is $10^{((20+\log a)/11)}$, which after simple manipulations is found to be equal to $100 \sqrt[11]{a/100}$. Now, it is obvious that the values close to 0 can take the output close to 0, whereas the values close to 255 will result in values only slightly away from 100. From this we can conclude that the geometric mean is more efficient in removing positive impulses than negative impulses, especially if the impulsive values are close to the extremal values. Similar analysis shows also that the nonlinear mean filters $CONTR_{-p}$, L_{-p}, and HARM have the same property, whereas the nonlinear mean filters L_p and $CONTR_p$ remove negative impulses better than positive impulses. To make the nonlinear filter able to remove both-sided noise, e.g., a cascade form, L_p mean followed by the L_{-p} mean has been proposed [88]. The result of this operation can easily be unsatisfactory as the first filter modifies "wrong" sided impulses by spreading them to their surroundings. The second operation cannot remove the resulting blotches and often makes them even more disturbing.

The geometric mean filter blurs step edges in a similar way to the arithmetic mean filter. The other nonlinear mean filters do better (but do not preserve edges completely), although they shift the edges slightly. The higher the value of p, the better the edge preservation.

3.14.3   Filtering Examples

From the one-dimensional examples it is clearly seen that these nonlinear mean filters remove positive impulses better than negative ones.

Since the nonlinear mean filters fail to remove two-sided impulses, we show the results using, in the noisy "Leena", only one-sided impulses. The new noisy "Leena" is shown in Figure 3.113.

The order (3.19) of the outputs of the nonlinear filters is clearly seen in the examples, especially in the two-dimensional images. The blurring effects are clear as well. Furthermore, the problems of the cascade form L_p mean followed by the L_{-p} mean are evident.

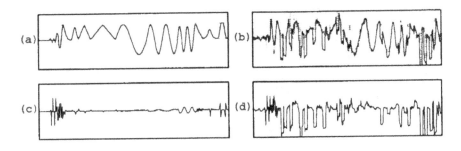

Figure 3.103. One-dimensional signals filtered by the geometric mean filter of window length 11.

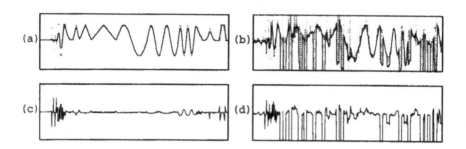

Figure 3.104. One-dimensional signals filtered by the harmonic mean filter of window length 11.

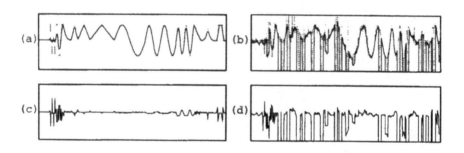

Figure 3.105. One-dimensional signals filtered by the L_p mean filter of window length 11, $p = -3$.

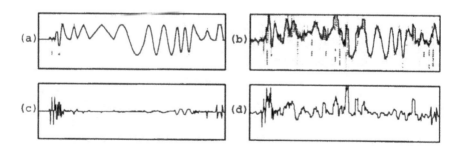

Figure 3.106. One-dimensional signals filtered by the L_p mean filter of window length 11, $p = 3$.

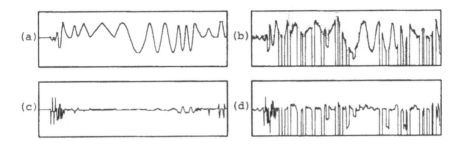

Figure 3.107. One-dimensional signals filtered by the contraharmonic mean filter of window length 11, $p = -3$.

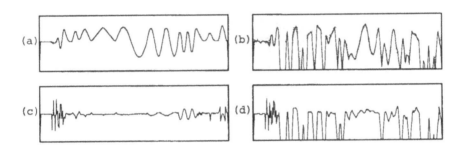

Figure 3.108. One-dimensional signals filtered by the L_p mean filter followed by the L_{-p} mean filter of window length 11, $p = -1.2$.

Figure 3.109. "Geometrical" filtered by the geometric mean filter of window size 5 × 5.

Figure 3.110. "Geometrical" filtered by the harmonic mean filter of window size 5×5.

Figure 3.111. "Geometrical" filtered by the L_p mean filter of window size 5×5, $p = -3$.

Figure 3.112. "Geometrical" filtered by the contraharmonic mean filter of window size 5×5, $p = -3$.

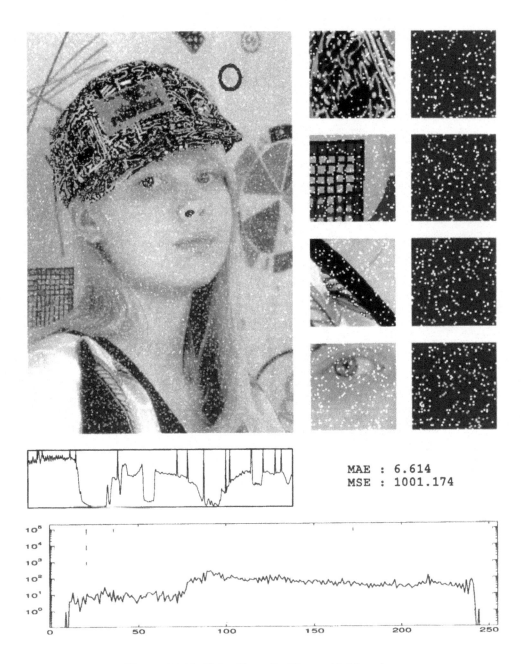

MAE : 6.614
MSE : 1001.174

Figure 3.113. Noisy "Leena" with one-sided impulses.

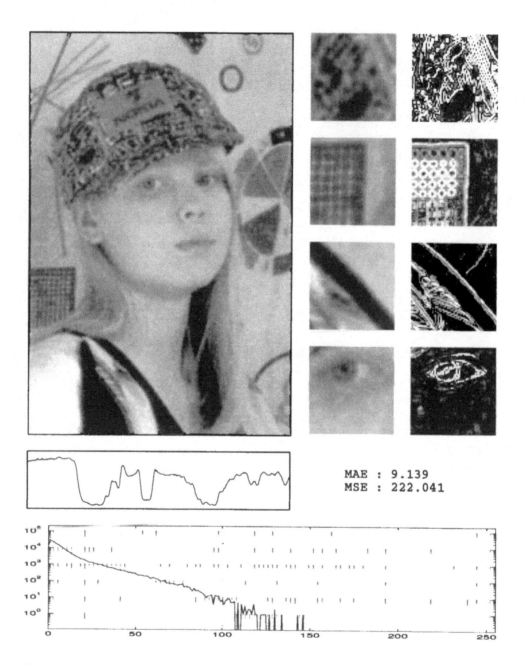

MAE : 9.139
MSE : 222.041

Figure 3.114. "Leena" (Figure 3.113) filtered by the geometric mean filter of window size 5 × 5.

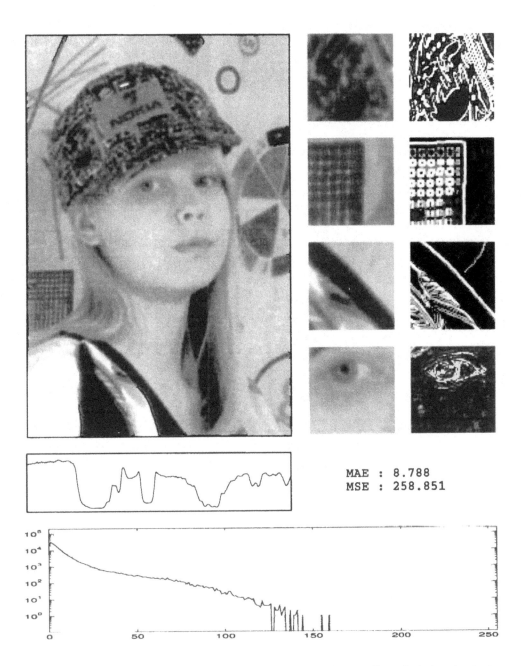

MAE : 8.788
MSE : 258.851

Figure 3.115. "Leena" (Figure 3.113) filtered by the harmonic mean filter of window size 5 × 5.

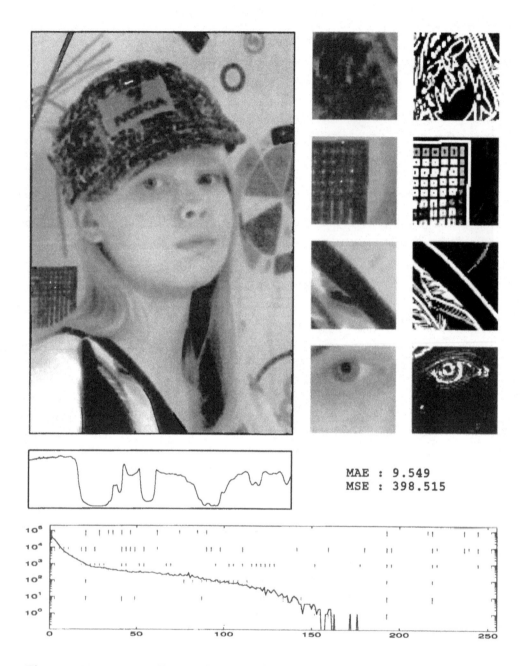

MAE : 9.549
MSE : 398.515

Figure 3.116. "Leena" (Figure 3.113) filtered by the L_p mean filter of window size 5×5, $p = -3$.

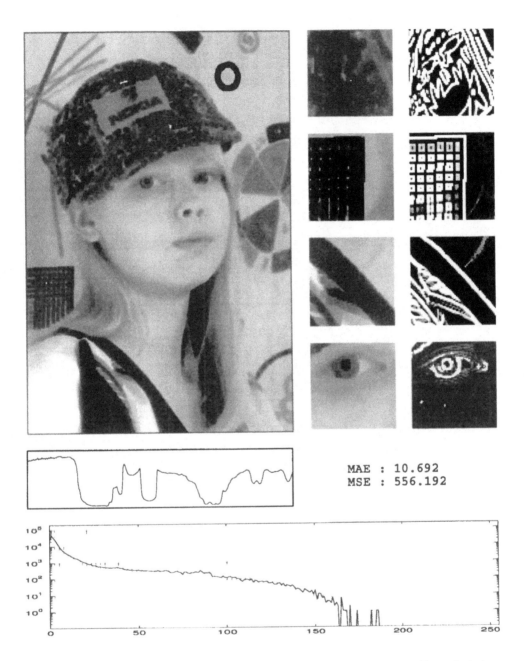

MAE : 10.692
MSE : 556.192

Figure 3.117. "Leena" (Figure 3.113) filtered by the contraharmonic mean filter of window size 5×5, $p = -3$.

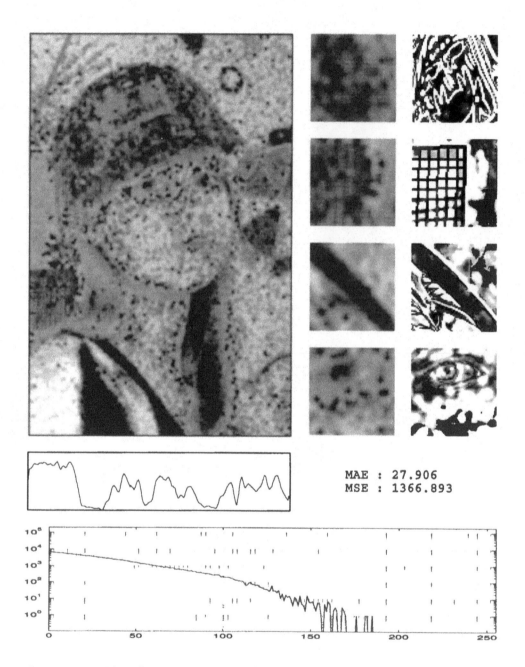

MAE : 27.906
MSE : 1366.893

Figure 3.118. "Leena" (Figure 1.15) filtered by the L_p mean filter followed by the L_{-p} mean filter of window sizes 5×5, $p = 1.2$.

3.15 Stack Filters

3.15.1 Principles and Properties

Binary signals are easy to manipulate and analyze. For example, the computation of the median value of a set of binary samples is trivial. If the sum of the samples in a window of size $N = 2k + 1$ is larger than k, then the median is one; otherwise it is zero. Thus, the ordering operation for binary signals reduces to additions. This observation can be exploited when designing digital nonlinear filters.

Consider an M-valued signal $x(i)$ with values from the set $\{0, 1, \ldots, M - 1\}$. Let us take slices from the signal on the level j in the following way: if the value of the signal is greater than or equal to j, let the value of the slice in the same position be one; otherwise let it be zero. The resulting slice $x^{(j)}(i)$ is then a binary signal obtained by the following *threshold decomposition function*

$$x^{(j)}(i) = T^{(j)}(x(i)) = \begin{cases} 1, & \text{if } x(i) \geq j, \\ 0, & \text{otherwise.} \end{cases} \tag{3.20}$$

The slices on the levels $1, 2, \ldots, M$ form simply a decomposition of the signal $x(i)$ into $M - 1$ binary signals, where all the information about the signal is preserved. The original signal can be reconstructed from its binary slices (see Figure 3.119)

$$x(i) = \sum_{m=1}^{M-1} x^{(m)}(i).$$

Consider median filtering of the obtained slice. The resulting signal is still binary and has the value 1 in the positions where the filter window has more ones than zeros, and has the value 0 otherwise. What does the number of ones, say n, in the filter window placed in the position k tell us? It surely tells that in this place in the filter window there are exactly n samples having their values greater than or equal to j. Thus, we notice that the value in the position k of the filtered binary slice is 1 if and only if the median of the original M-valued signal in the same position is greater than or equal to j. In other words, the result of the median filter in the original signal defines the values of the filtered binary slices in the same way that the original signal defines the values of the binary slices. See Figure 3.119.

The median filtered signal can be reconstructed from its binary slices, which can be obtained in two different ways:

1. Use threshold decomposition function to the median filtered signal.

2. Use threshold decomposition function to the original signal and filter the resulting binary slices by the median filter.

The second way provides us a method of implementing the median filter by composing the original M-valued signal to $M - 1$ binary signals, filtering these signals independently and combining the results by summing to obtain the median filter output. This method is depicted in Figure 3.119, where we have used a three-point median filter as an example.

Now, the following idea may come to one's mind: what happens if we change the binary function, i.e., use some other function operating on binary values instead of the binary median function? This idea leads to the definition of *stack filters*. First, some

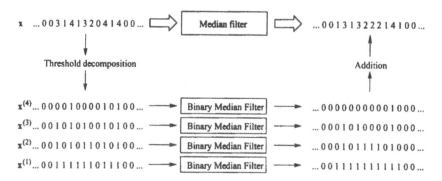

Figure 3.119. Illustration of the median filtering ($N = 3$) operation using threshold decomposition. The broad arrows show the overall filtering operation. The slender arrows show the same operation in the threshold decomposition architecture.

definitions are needed.

Let $\mathbf{x}$ and $\mathbf{y}$ be binary vectors of length n. Define

$$\mathbf{x} \leq \mathbf{y} \text{ if and only if } x_n \leq y_n \text{ for all } n. \tag{3.21}$$

Since the relation defined by (3.21) is reflexive, antisymmetric, and transitive, it defines a partial ordering on the set of binary vectors of fixed length. Now, consider a signal $\mathbf{x}$ and its thresholded binary signals $\mathbf{x}^{(1)}, \mathbf{x}^{(2)}, \ldots, \mathbf{x}^{(M-1)}$. Clearly,

$$\mathbf{x}^{(i)} \leq \mathbf{x}^{(j)} \text{ if } i \geq j.$$

Thus, the binary signals $\mathbf{x}^{(1)}, \mathbf{x}^{(2)}, \ldots, \mathbf{x}^{(M-1)}$ are non-increasing in the sense of the partial ordering of (3.21).

It has turned out that by defining filtering operations based on those binary operators (Boolean functions) $f(\cdot)$ for which it holds for any $\mathbf{x}$ that

$$f(\mathbf{x}^{(i)}) \leq f(\mathbf{x}^{(j)}) \text{ if } i \geq j \tag{3.22}$$

we obtain a class of filters with many useful properties. The condition (3.22) basically ensures that there will be no "holes" in the filtered stack of binary signals, i.e., the filtered stack directly corresponds to a threshold decomposed signal. This requirement has led to the definition of stack filters based on positive Boolean functions [111, 112].

A Boolean function $f(\cdot)$ is called a *positive Boolean function* if it can be written as a Boolean expression that contains only uncomplemented input variables. For a positive Boolean function $f(\cdot)$ it holds that

$$f(\mathbf{x}) \geq f(\mathbf{y}) \text{ if } \mathbf{x} \geq \mathbf{y}. \tag{3.23}$$

The property (3.23) is called the *stacking property*. A complete proof that (3.23) holds for positive Boolean functions can be found in Reference [77]. As (3.23) holds for positive Boolean functions, (3.22) holds for them as required.

Remark 3.1. We use the multiplication $\cdot$ to denote conjunction and the addition $+$ to denote disjunction in the Boolean functions. Thus, $x_1 \vee (x_2 \wedge x_3)$ is in our notation $x_1 + x_2 x_3$. The complement is denoted by the overbar, i.e., $\overline{x_1}$ denotes the complement of x_1.

Example 3.29. The Boolean function $f(x_1, x_2, x_3) = x_1 x_2 + \overline{x_1} x_2 + x_3$ is a positive Boolean function, as it can be written as $x_2(x_1 + \overline{x_1}) + x_3 = x_2 + x_3$. Now, one can verify that (3.23) holds by going through all vectors $\mathbf{x}$ and $\mathbf{y}$ of length 3 satisfying $\mathbf{x} \geq \mathbf{y}$. For example, $\mathbf{x} = (0, 1, 1) \geq \mathbf{y} = (0, 1, 0)$ and $f(\mathbf{x}) = 1 \geq f(\mathbf{y}) = 1$.

One can always represent a positive Boolean function in the *minimum sum-of-products form*

$$f(x_1, x_2, \ldots, x_N) = \sum_{i=1}^{K} \prod_{j \in P_i} x_j,$$

where P_i are subsets of $\{1, 2, \ldots, N\}$. This representation is unique.

A *stack filter* is defined by a positive Boolean function $f(\cdot)$ as follows:

$$\text{Stack}(X_1, X_2, \ldots, X_N; f(\cdot)) = \sum_{m=1}^{M-1} f(\mathbf{x}^{(m)}). \tag{3.24}$$

Thus, filtering a vector $\mathbf{x}$ by a stack filter $\text{Stack}(\cdot; f(\cdot))$ based on the positive Boolean function $f(\cdot)$ is equivalent to decomposing $\mathbf{x}$ to binary vectors $\mathbf{x}^{(m)}$, $1 \leq m \leq M - 1$, by thresholding, then filtering each threshold level by the binary filter $f(\cdot)$ and reconstructing the output vector as the sum (3.24).

By (3.24), stack filters are completely characterized by their operation on binary vectors. The importance of this property arises from the fact that binary vectors are easier to analyze than multivalued vectors. Also, filtering each binary vector independently allows the operation to be done in parallel, and single binary filters are easy to implement.

Stack filters can also be implemented without using the threshold decomposition. Let $\mathbf{x} = (X_1, X_2, \ldots, X_N)$ be an input vector to a stack filter $S_f(\cdot)$ defined by a positive Boolean function $f(x_1, x_2, \ldots, x_N)$ and P_i be subsets of $\{1, 2, \ldots, N\}$. Then

$$f(x_1, x_2, \ldots, x_N) = \sum_{i=1}^{K} \prod_{j \in P_i} x_j,$$

if and only if the stack filter $\text{Stack}(\cdot)$ corresponding to $f(x_1, x_2, \ldots, x_N)$ is [98]

$$\text{Stack}(\mathbf{x}; f(\cdot)) = \text{MAX}\{\text{MIN}\{X_j : j \in P_1\}, \text{MIN}\{X_j : j \in P_2\}, \ldots, \text{MIN}\{X_j : j \in P_K\}\}.$$

Thus, the real domain stack filter corresponding to a positive Boolean function can be expressed by replacing AND and OR with MIN and MAX, respectively. For example, the three point median filter over real variables X_1, X_2, and X_3 is a stack filter defined by the positive Boolean function $f(x_1, x_2, x_3) = x_1 x_2 + x_1 x_3 + x_2 x_3$, i.e.,

$$\text{MED}\{X_1, X_2, X_3\} = \text{MAX}\{\text{MIN}\{X_1, X_2\}, \text{MIN}\{X_1, X_3\}, \text{MIN}\{X_2, X_3\}\}.$$

An algorithm for stack filtering based on this max-min realization is shown in Algorithm 3.26. The sets P_i are placed into the code as subwindows. Thus, in every subfilter the output is the minimum value inside the subwindow. The output of the stack filter is the maximal value of these outputs.

Example 3.30. Let the 5-point input be $\mathbf{x} = (2, 7, 9, 20, 5)$ and the Boolean function be $f(x_1, x_2, x_3, x_4, x_5) = x_1 x_2 + x_2 x_3 x_4 + x_4 x_5$. The output of the stack filter defined by $f(x_1, x_2, x_3, x_4, x_5)$ is $\text{MAX}\{\text{MIN}\{X_1, X_2\}, \text{MIN}\{X_2, X_3, X_4\}, \text{MIN}\{X_4, X_5\}\} = \text{MAX}\{2, 7, 5\} = 7$.

Stack filter

Inputs: *NumberOfRows* × *NumberOfColumns* image
　　　　　Set of subwindows $\{P_1, P_2, \ldots, P_K\}$
Output: *NumberOfRows* × *NumberOfColumns* image

```
for i = 1 to NumberOfRows
    for j = 1 to NumberOfColumns
        let FoundMax = SmallestPossibleValue
        for l = 1 to K
            place the subwindow P_l at (i,j)
            find the minimum Min of the values of the image inside P_l
            if Min > FoundMax
                let FoundMax = Min
            end
        end
        let Output(i,j) = FoundMax
    end
end
```

Algorithm 3.26. Algorithm for the stack filter.

Out of the filters considered so far, all weighted order statistic, ranked-order (and their subclasses: weighted median and median), max/median, and multistage median filters are also stack filters (see Exercises). In the next sections we will consider some more subclasses of stack filters. So, the class of stack filters is very large and covers a wide variety of filters.

3.15.2　Impulse and Step Response

Consider only nontrivial Boolean functions, i.e., functions not identically equal to one or zero. The impulse response of a stack filter can be easily obtained from its maximum of minima representation. If a negative impulse is to be the output for some window position, it should be the largest of the minima. This can happen only if all the minima have the same value, i.e., that of the impulse. But this would require that one sample X_i must belong to every P_j. Furthermore it means that the positive Boolean function must be of the form $x_i \cdot g(\mathbf{x})$, where $g(\mathbf{x})$ is a Boolean function. A positive impulse will be the output if and only if it is one of the minima. This is equivalent to the existence of a set P_j with only one element X_k, or equivalently, the positive Boolean function is of the form $x_k + g(\mathbf{x})$, where $g(\mathbf{x})$ is a Boolean function. In all other cases, stack filters remove impulses completely.

Consider the case where a step edge enters a stack filter. This is equivalent to a binary step edge entering the binary filter. The signal segments inside the window are

$$(0, \ldots, 0), (0, \ldots, 0, 1), (0, \ldots, 0, 1, 1), \ldots, (0, 1, \ldots, 1), (1, \ldots, 1).$$

These binary vectors form an increasing sequence and thus for any positive Boolean function the outputs form the sequence

$$0 = f(0,\ldots,0),\ldots,0 = f(0,\ldots,0,\overset{s}{\underset{\downarrow}{1}},\ldots,1), 1 = f(0,\ldots,0,\overset{s-1}{\underset{\downarrow}{1}},\ldots,1),\ldots,1 = f(1,\ldots,1).$$

Thus, any stack filter preserves an ideal edge but may cause edge shift.

3.15.3 Filtering Examples

The Boolean function used in the one-dimensional signal is $x_1 x_2 x_3 x_4 x_5 + x_6(x_3 x_{10} + x_3 x_9 + x_4 x_8 + x_5 x_7) + x_7 x_8 x_9 x_{10} x_{11}$. In the two-dimensional examples the filter is

$$\text{MAX}\{\text{MED}\{X_i : i \in Q\}, \text{MIN}\{X_j : j \in P_1\}, \text{MIN}\{X_j : j \in P_2\}, \ldots, \text{MIN}\{X_j : j \in P_8\}\},$$

where Q is the 3×3 square shown in the right hand side of Figure 3.54 and the sets P_i are the sets defining the other eight subwindows in Hachimura-Kuwahara filtering; see Figure 3.54. Other examples of stack filtering can be seen from other sections where the considered filters form subsets of stack filters. These filters have been mentioned above.

The filters used are relatively efficient in noise removal but are also able to preserve details simultaneously.

3.16 Generalizations of Stack Filters

3.16.1 Principles and Properties

Stack filters are a subclass of a class of so called *generalized stack filters* [68]. This class of filters is the set of all filters possessing the above threshold decomposition architecture and the stacking property.

In stack filters the binary vector obtained by thresholding the input vector at level l is fed into a positive Boolean function, which is the same for all levels. In generalized stack filters, any number of threshold vectors can be fed into the filter at level l. A generalized stack filter uses a binary array as its input, instead of just a binary vector, and different Boolean functions are allowed to operate at each threshold level.

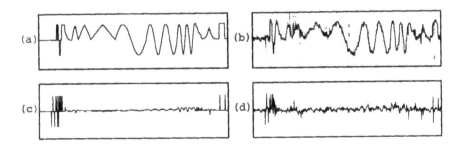

Figure 3.120. One-dimensional signals filtered by the stack filter of window length 11; the Boolean function is given in the text.

Figure 3.121. "Geometrical" filtered by the stack filter of window size 5×5; the Boolean function is given in the text.

Let us define the *binary threshold array* of an input $\mathbf{x} = (X_1, X_2, \ldots, X_N)$ by

$$
\mathbf{X}_m^I = \begin{pmatrix} \mathbf{x}^{(m+I)} \\ \vdots \\ \mathbf{x}^{(m)} \\ \vdots \\ \mathbf{x}^{(m-I)} \end{pmatrix} = \begin{pmatrix} x_1^{(m+I)} & x_2^{(m+I)} & \cdots & x_N^{(m+I)} \\ x_1^{(m+I-1)} & x_2^{(m+I-1)} & \cdots & x_N^{(m+I-1)} \\ \vdots & \vdots & \ddots & \vdots \\ x_1^{(m-I)} & x_2^{(m-I)} & \cdots & x_N^{(m-I)} \end{pmatrix},
$$

where $I \in \{0, 1, \ldots, M-1\}$ and $m \in \{1, 2, \ldots, M-1\}$.

In order to account for all binary threshold arrays we define the "extra" binary threshold vectors as follows:

$$
\mathbf{x}^{(n)} = \begin{cases} \mathbf{0}^T, & \text{if } n > M-1 \\ \mathbf{1}^T, & \text{if } n < 1, \end{cases}
$$

where $\mathbf{0}$ and $\mathbf{1}$ are column vectors of length N in which every entry is 0 or 1, respectively.

Thus, $\mathbf{X}_m^I$ is a $(2I+1) \times N$ binary array obtained from $\mathbf{x}$, where each row of $\mathbf{X}_m^I$ is a particular binary threshold vector of $\mathbf{x}$. Each binary threshold array possesses the internal stacking property, since

$$
\mathbf{x}^{(k)} \geq \mathbf{x}^{(l)}, \text{ whenever } l \geq k.
$$

The same reasoning shows that the binary threshold arrays also obey the stacking property, i.e.,

$$
\mathbf{X}_1^I \geq \mathbf{X}_2^I \geq \cdots \geq \mathbf{X}_{M-1}^I.
$$

The ordered set of $M-1$ Boolean functions $\{g_1(\cdot), g_2(\cdot), \ldots, g_{M-1}(\cdot)\}$ is called a *stacking set of Boolean functions* if

$$
g_m(\mathbf{X}_m^I) \geq g_{m+1}(\mathbf{X}_{m+1}^I), \ 1 \leq m \leq M-2.
$$

The generalized stack filter GenStack$(\cdot; I, G)$ is defined by a stacking set of $M-1$ Boolean functions $G = \{g_1(\cdot), g_2(\cdot), \ldots, g_{M-1}(\cdot)\}$, where the input to the Boolean function of level m is the binary threshold array $\mathbf{X}_m^I$. The operation of this filter is defined as follows:

$$
\text{GenStack}(\mathbf{x}; I, G) = \sum_{m=1}^{M-1} g_m(\mathbf{X}_m^I).
$$

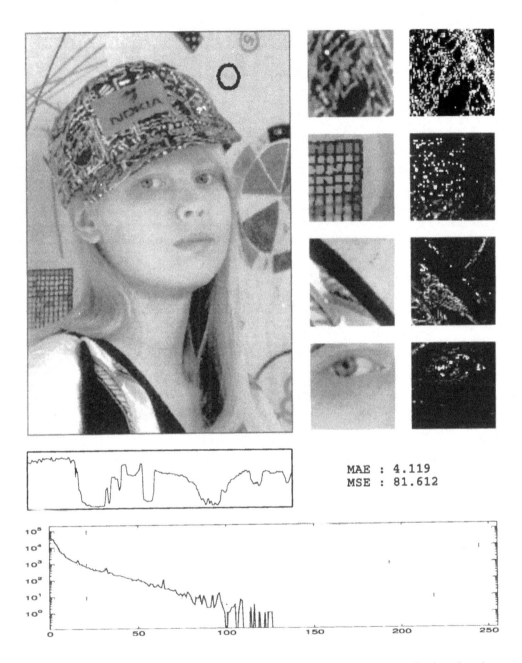

MAE : 4.119
MSE : 81.612

Figure 3.122. "Leena" filtered by the stack filter of window size 5 × 5; the Boolean function is given in the text.

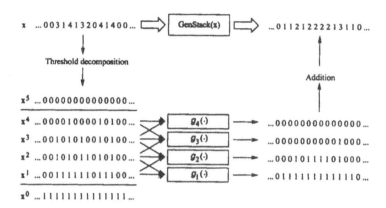

Figure 3.123. A generalized stack filter $S_1(\cdot)$ with window width $N = 3$ and with $2I + 1 = 3$ binary threshold vectors fed into each Boolean filter in the thresholding architecture of $S_1(\cdot)$. The Boolean functions are shown in (3.25). The multilevel operation of the filter is shown by the broad arrows. The slender arrows show the same operation in the threshold decomposition architecture.

The input to the Boolean function $g_m(\cdot)$ at level m is the binary threshold array $\mathbf{X}_m^I$, which $g_m(\cdot)$ maps into a single bit. The bits are then summed to obtain the M-valued output of the filter. This is illustrated in Figure 3.123, where the Boolean functions are defined as follows: Let the indices of the variables of the binary threshold arrays be

$$\mathbf{X} = \begin{pmatrix} x_{1,-1} & x_{1,0} & x_{1,1} \\ x_{0,-1} & x_{0,0} & x_{0,1} \\ x_{-1,-1} & x_{-1,0} & x_{-1,1} \end{pmatrix}$$

and the Boolean functions be

$$\begin{aligned}
g_1(\mathbf{X}) &= x_{0,-1} + x_{0,0} + x_{0,1}; \\
g_2(\mathbf{X}) &= x_{0,-1}x_{0,0} + x_{0,-1}x_{0,1} + x_{0,0}x_{0,1}; \\
g_3(\mathbf{X}) &= (x_{0,-1}x_{0,0} + x_{0,-1}x_{0,1} + x_{0,0}x_{0,1})(x_{1,-1}x_{1,0} + x_{1,-1}x_{1,1} + x_{1,0}x_{1,1}); \\
g_4(\mathbf{X}) &= x_{0,-1}x_{0,0}x_{0,1}x_{-1,-1}x_{-1,0}x_{-1,1}x_{1,-1}x_{1,0}x_{1,1}.
\end{aligned} \tag{3.25}$$

The threshold decomposition architecture and the requirement of the stacking property lead to an interesting interpretation of the operation of the filter at each level. The filter at level m decides whether the desired signal is less than m or not. The decision is based on the corrupted observations available from the binary threshold array input. The stacking property ensures that the decisions on different levels of the architecture are consistent. In other words, if the filter at level i decides that the signal is less than i at the current time, then all filters at levels $i + 1$ to $M - 1$ must draw the same conclusion. A necessary and sufficient condition for this consistency is the use of a stacking set of Boolean functions for the filters at the various levels.

A generalized stack filter is said to be *homogeneous* if it has the same Boolean function at each level in the architecture; otherwise, it is said to be *inhomogeneous*.

The question of whether or not all functions in the stacking set need to be positive is of particular interest. The answer is no, since, for example, the functions

$$g_4(\mathbf{x}) = x_2 x_3 x_4 (\overline{x}_1 + \overline{x}_5);$$
$$g_3(\mathbf{x}) = x_2 x_3 x_4;$$
$$g_2(\mathbf{x}) = (x_2 x_3 + x_3 x_4)(\overline{x}_1 + \overline{x}_5) + x_1 x_3 x_5;$$
$$g_1(\mathbf{x}) = x_1 + x_2 + x_3 + x_4 + x_5$$

(3.26)

form a stacking set for a window of width 5 and $M = 5$ (Exercise). Thus, for an inhomogeneous generalized stack filter with $I = 0$ the set of Boolean functions defining the filter need not be positive. The same statement can also be made when $I \neq 0$.

Example 3.31. Consider the 5-point input $\mathbf{x} = (2, 1, 4, 3, 0)$, $M = 5$, and the stacking set given in (3.26). The thresholded signals and the results after processing by the Boolean functions in the stacking set are

$$\mathbf{x}^4 = (0, 0, 1, 0, 0) \quad g_4(\mathbf{x}^4) = 0$$
$$\mathbf{x}^3 = (0, 0, 1, 1, 0) \quad g_3(\mathbf{x}^3) = 0$$
$$\mathbf{x}^2 = (1, 0, 1, 1, 0) \quad g_2(\mathbf{x}^2) = 1$$
$$\mathbf{x}^1 = (1, 1, 1, 1, 0) \quad g_1(\mathbf{x}^1) = 1.$$

By summing the results we obtain the output, which is 2.

Threshold Boolean filters are also natural extensions of stack filters [9, 62]. In stack filtering the binary domain filter is assumed to be a positive Boolean function. If this requirement is relaxed so that the binary domain filter can be any Boolean function, new filters are found and Threshold Boolean filters are obtained (see Figure 3.124). The positivity of the Boolean function in fact implies that any stack filter is an increasing operation, i.e., if $f(x)$ and $g(x)$ are signals where $f(x) \leq g(x)$ for all x, then $\mathrm{Stack}(f(x)) \leq \mathrm{Stack}(g(x))$ for all x. This is a very strong condition and makes the class of stack filters fairly tractable. It has turned out, especially in document processing, that the noise characteristics are typically such that increasing filters do not provide a satisfactory solution. The difficulties follow from the fact that the noise is usually far from uncorrelated i.i.d. noise, and it is also quite signal dependent.

Example 3.32. Consider the 5-point input $\mathbf{x} = (2, 1, 4, 3, 0)$, $M = 5$, and the Boolean function $f(\mathbf{x}) = x_2 x_3 x_4 + \overline{x}_1 \overline{x}_5$. The thresholded signals and the results after processing by the Boolean functions are now

$$\mathbf{x}^4 = (0, 0, 1, 0, 0), \quad f(\mathbf{x}^4) = 1$$
$$\mathbf{x}^3 = (0, 0, 1, 1, 0), \quad f(\mathbf{x}^3) = 1$$
$$\mathbf{x}^2 = (1, 0, 1, 1, 0), \quad f(\mathbf{x}^2) = 0$$
$$\mathbf{x}^1 = (1, 1, 1, 1, 0), \quad f(\mathbf{x}^1) = 1.$$

By summing the results we obtain the output, which is 3.

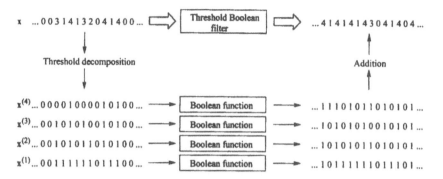

Figure 3.124. Illustration of the threshold Boolean filtering ($N = 3$) operation using threshold decomposition. The broad arrows show the overall filtering operation. The slender arrows show the same operation in the threshold decomposition architecture. The Boolean function used in this illustration is $x_2 + \bar{x}_1 \bar{x}_3$.

3.16.2   Impulse and Step Response

Generalized stack filters can remove impulses, change their magnitude, or spread them. Step edges can also be modified in various ways.

Almost all stack filters remove impulses. Threshold Boolean filters in turn can have a quite interesting impulse response. Consider filtering the binary signal $\{\ldots, 1, 1, 0, 1, 1 \ldots\}$ by the threshold Boolean filter resulting from $\bar{x}_1 x_2$. The corresponding real domain threshold Boolean filter can be shown to be $\mathrm{MAX}\{0, X_2 - X_1\}$ (Exercise). The result is then the signal $\{\ldots, 0, 0, 1, 0, 0, \ldots\}$. In other words, a negative impulse is changed into a positive impulse! We saw that any stack filter is basically a smoother with unity gain so that it will always preserve edges (possibly shifting them). On the other hand, threshold Boolean filters can be designed to perform as edge detectors. The simplest way is to utilize the edge shifting property of stack filters by combining two stack filters with a different edge shift [9]. This leads to noise-suppressing edge detectors. For example, consider the Boolean function $x_1 \oplus x_2$, where $\oplus$ denotes modulo-2 operation. This corresponds to $|X_1 - X_2|$ in the real domain (Exercise). The step edges $\{\ldots, a, a, a, b, b, \ldots\}$ will be changed into $\{\ldots, 0, 0, |a - b|, 0, 0, \ldots\}$, and the edge location is marked.

3.16.3   Filtering Examples

Again for practical reasons we avoid showing examples. This would have required showing dozens of lengthy Boolean functions, and in our opinion it requires quite a deep understanding of generalized stack (threshold Boolean) filtering to extract information relating the functions and the results together.

3.17   Morphological Filters

3.17.1   Principles and Properties

Morphological filters are nonlinear filters based on morphological transformations of signals by sets. Originally *mathematical morphology*, developed by Serra [95] and Matheron

[75], has been used for the description of binary images and objects. The purpose of mathematical morphology is the quantitative description of geometrical structures of signals. Nowadays, mathematical morphology is an important discipline and is widely used in various signal processing tasks. Certain morphological transformations behave like filters and may be used for solving the problems discussed in Chapter 1. Thus, morphological filters do not actually originate from these problems, but are based on an existing solid theory which is primarily used in other signal processing tasks. Refer to Lantuéjoul and Serra [58],

> Mathematical morphology provides an alternative to classical filtering methods, such as convolution. To replace linearity by the increasing condition leads to the class of the *M*-filters. Although they are not invertible, they remove very accurately specific features from the images (peaks but not holes for ex.).

Thus, originally Lantuéjoul and Serra used the name *M-filters* for morphological filters [58]. A filter $F(\cdot)$ is called morphological if it is

- *increasing*: $f(x) \leq g(x)$ for all x implies that $F(f(x)) \leq F(g(x))$ for all x;

- *idempotent*: $F(F(f(x))) = F(f(x))$ holds for all f and x.

The increasing condition means that the filter is order preserving. Idempotence guarantees that any signal is changed by the filter into a *root signal* by one filtering iteration (about root signals, see Exercises). In other words, the repeated filtering does not change the signal any more.

The moving window of morphological filters is typically called the *structuring element*. Let us denote the domain of the structuring element by $B \subset \mathbf{R}^n$. The domain describes the neighborhood by defining the shape of the moving window. The symmetric set B^s of B is given by

$$B^s = \{-x : x \in B\}.$$

In other words, it is obtained by rotating B 180 degrees in the plane.

Every element in the structuring element is assigned a weight, a real number $B_i : i \in B$.

In morphological filtering, the geometrical features of a signal are modified by processing the signal by a structuring element. The choice of the structuring element is made based on its geometrical properties. The basic morphological operations are *dilation* and *erosion*. These operations are not considered to be morphological filters at all as they clearly are not idempotent. However, we also consider these to be nonlinear filters. The main problem associated with dilation and erosion is their limited possible usage in real signal processing applications.

The idea behind morphological filters is more easily understood by considering *flat structuring elements* in the stack filter framework. For flat structuring elements the weights are all zero, $B_i = 0, i \in B$.

The dilation by the flat structuring element B is given by

$$\text{Dil}(X_1, X_2, \ldots, X_N) = \text{MAX}\{X_1, X_2, \ldots, X_N\},$$

and the erosion by the structuring element B is given by

$$\text{Ero}(X_1, X_2, \ldots, X_N) = \text{MIN}\{X_1, X_2, \ldots, X_N\}.$$

Thus, the flat dilation and the flat erosion are simply the Nth and 1st ranked-order filters, respectively, with the moving window B.

The dilation (erosion) of a signal $F(\cdot)$ by a structuring element B is often denoted by $F \oplus B^s$ ($F \ominus B^s$). This notation may seem to be a little confusing. The objective of this notation is to make a distinction between the dilation (erosion) and the so-called Minkowski set addition (subtraction). Some authors do not make this distinction, leading to slightly different notation.

Geometrically dilations can be understood as operations expanding the original images, and erosions as operations shrinking the images.

Based on the dilation and the erosion, four compound operations are defined: the *closing*, the *opening*, the *close-opening*, and the *open-closing*.

$$
\begin{aligned}
F^B &= (F \oplus B^s) \ominus B; \\
F_B &= (F \ominus B^s) \oplus B; \\
(F^B)_B &= (((F \oplus B^s) \ominus B) \ominus B^s) \oplus B; \\
(F_B)^B &= (((F \ominus B^s) \oplus B) \oplus B^s) \ominus B.
\end{aligned}
$$

In other words:

- The closing of a signal F by B is defined as the dilation by B followed by the erosion by B^s;

- The opening of a signal F by B is defined as the erosion by B followed by the dilation by B^s;

- The close-opening of a signal F by B is defined as the closing by B followed by the opening by B;

- The open-closing of a signal F by B is defined as the opening by B followed by the closing by B.

These operations are increasing and idempotent, and thus they are morphological filters.

By the definition, flat morphological filters are sequences of maximum and minimum operations, which are stack filters. This means that flat morphological filters can be understood as stack filters originating from mathematical morphology.

To distinguish flat and other structuring elements the latter ones are called *gray-level structuring elements*. The dilation by the structuring element B is given by

$$\text{Dil}(X_1, X_2, \ldots, X_N; B) = \text{MAX}\{X_1 + B_1, X_2 + B_2, \ldots, X_N + B_N\}$$

and the erosion by the structuring element B is given by

$$\text{Ero}(X_1, X_2, \ldots, X_N; B) = \text{MIN}\{X_1 - B_1, X_2 - B_2, \ldots, X_N - B_N\}.$$

These operations clearly resemble linear convolution. The closing, opening, close-opening, and open-closing are defined in an analogous manner as above.

Closings are *extensive* operations, meaning that $F^B(f(x)) \geq f(x)$ for every signal $f(\cdot)$ and every x. Openings in turn are *antiextensive*, i.e., $F^B(f(x)) \leq f(x)$ for every signal $f(\cdot)$ and every x.

Morphological closing

Inputs: *NumberOfRows* × *NumberOfColumns* image
Structuring set B, $|B| = N = 2k + 1$
Weight vector of the structuring element $(B_1, B_2, \ldots, B_N)$
Output: *NumberOfRows* × *NumberOfColumns* image

```
/* Dilation by B */
for i = 1 to NumberOfRows
    for j = 1 to NumberOfColumns
        place the structuring set B at (i, j)
        for samples inside B, store the sum of the image samples and the
            corresponding weights in x = (X₁ + B₁, X₂ + B₂, ..., Xₙ + Bₙ)
        find the largest element, Max, of x
        let IntermediaryOutput(i, j) = Max
    end
end
/* Form symmetric structuring set Bˢ */
initialize Bˢ as an empty set
for every element (iₓ, jₓ) of B
    add (−iₓ, −jₓ) to Bˢ
end
/* Erosion by Bˢ */
for i = 1 to NumberOfRows
    for j = 1 to NumberOfColumns
        place the (symmetric) structuring set Bˢ at (i, j)
        for samples inside Bˢ, store the sum of the IntermediaryOutput image
            samples and the corresponding weights in
            x = (X'₁ − Bˢ₁, X'₂ − Bˢ₂, ..., X'ₙ − Bˢₙ)
        find the smallest element, Min, of x
        let Output(i, j) = Min
    end
end
```

Algorithm 3.27. Algorithm for the closing.

We give the algorithm only for closing (see Algorithm 3.27); based on this, algorithms for the other operations can be easily derived.

Example 3.33. Let us choose the structuring element with weights $B_{-1} = B_0 = B_1 = B_2 = 0$. Note that this structuring element is flat but it is not symmetric around the origin; its domain is $B = \{-1, 0, 1, 2\}$ and $B^s = \{-2, -1, 0, 1\}$. Now, for the signal $\mathbf{x} = (3, 8, 5, 7, 2, 1, 1, 9, 8, 5)$ we obtain:

original	(3,	8,	5,	7,	2,	1,	1,	9,	8,	5);
dilated by B	(∗,	8,	8,	7,	7,	9,	9,	9,	∗,	∗);
dilated by B^s	(∗,	∗,	8,	8,	7,	7,	9,	9,	9,	∗);
eroded by B	(∗,	3,	2,	1,	1,	1,	1,	1,	∗,	∗);
eroded by B^s	(∗,	∗,	3,	2,	1,	1,	1,	1,	1,	∗).

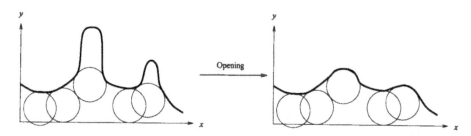

Figure 3.125. Rolling ball transformation

The erosion by B^s of the result after the dilation by B gives the closing:

dilated by B	(∗,	8,	8,	7,	7,	9,	9,	9,	∗,	∗);
closed by B	(∗,	∗,	∗,	7,	7,	7,	7,	∗,	∗,	∗).

Similarly we obtain the opening:

eroded by B	(∗,	3,	2,	1,	1,	1,	1,	1,	∗,	∗);
opened by B	(∗,	∗,	∗,	3,	2,	1,	1,	∗,	∗,	∗).

If we use a gray-level structuring element with the same domain, $B_{-1} = 3, B_0 = 6, B_1 = 3, B_2 = 1$, we obtain:

original	(3,	8,	5,	7,	2,	1,	1,	9,	8,	5);
dilated by B	(∗,	14,	11,	13,	10,	10,	12,	15,	∗,	∗);
dilated by B^s	(∗,	∗,	11,	13,	10,	8,	12,	15,	14,	∗);
eroded by B	(∗,	0,	−1,	−1,	−4,	−5,	−5,	−2,	∗,	∗);
eroded by B^s	(∗,	∗,	−1,	−1,	−4,	−5,	−5,	−2,	0,	∗);
closed by B	(∗,	∗,	∗,	7,	4,	4,	6,	∗,	∗,	∗);
opened by B	(∗,	∗,	∗,	5,	2,	1,	1,	∗,	∗,	∗).

Figure 3.125 gives a graphical illustration of the opening operation. The opening is understood as a rolling ball transformation, where the rolling ball traces the smoothly varying contours, deletes positive impulses and broadens the negative impulses. The closing is an opposite rolling ball operation, deleting negative and expanding positive impulses.

The *alternating sequential filters* form a particular class of morphological filters [35, 96]. Sternberg [98] studied noise removal by filtering the signal first by a small opening, then by a slightly larger closing, followed by a slightly larger opening, etc. This is what is called an alternating sequential filter: a composition of openings and closings with increasingly larger structuring elements. The main problem with the alternating sequential filters is to know when to stop filtering. If it is stopped after too few passes noise may not be sufficiently eliminated, and if it is continued up to too many passes the whole signal may be eliminated.

3.17.2   Impulse and Step Response

The opening and the closing can only remove one-sided impulses. The flat opening and closing also preserve steps completely, whereas openings and closings with gray-level structuring elements may slightly blur step edges (Exercise). The close-opening and the open-closing can remove both negative and positive impulses. As they are defined as compositions of opening and closing their step responses follow from that of the opening and closing.

3.17.3   Filtering Examples

The one-dimensional results are obtained by using flat structuring elements of length 5. This means that in the opening and the closing (the close-opening and the open-closing) altogether 4 (8) samples from both sides of the sample currently under filtering have an effect on the result. In the two-dimensional examples the flat structuring element is the 3×3 square with the origin as its center.

From the examples the extensivity of the closings and the antiextensivity of the openings are clearly seen. Furthermore, the fact that openings and closings can only remove one-sided impulses is well pronounced in the examples. The blurring effects of the operations are also evident.

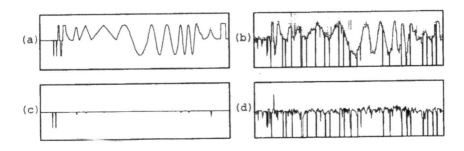

Figure 3.126. One-dimensional signals filtered by the flat opening of window length 5.

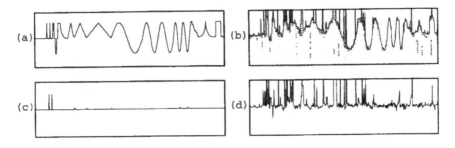

Figure 3.127. One-dimensional signals filtered by the flat closing of window length 5.

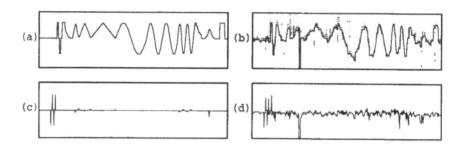

Figure 3.128. One-dimensional signals filtered by the flat open-closing of window length 5.

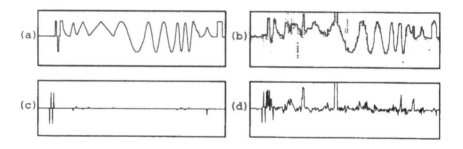

Figure 3.129. One-dimensional signals filtered by the flat close-opening of window length 5.

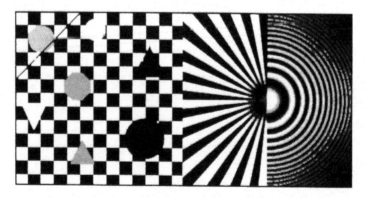

Figure 3.130. "Geometrical" filtered by the flat opening of window size 3×3.

Figure 3.131. "Geometrical" filtered by the flat closing of window size 3 × 3.

Figure 3.132. "Geometrical" filtered by the flat open-closing of window size 3 × 3.

3.18   Soft Morphological Filters

3.18.1   Principles and Properties

The idea behind the so-called *soft morphological filters* [51, 55] is to slightly relax the standard definitions of flat morphological filters in such a way that a degree of robustness is achieved, while most of the desirable properties of standard flat morphological operations are maintained. As discussed above, standard flat morphological filters are based on local maximum and minimum operations, while soft morphological filters are based on more general weighted order statistics. The main difference between soft and standard flat morphological filters is that soft morphological filters are less sensitive to additive noise and small variations in the shapes of the object to be filtered. What is typically lost by this relaxation is the idempotence property of the opening and the closing. So, strictly speaking, soft morphological filters are not morphological filters at all. Still the name "soft morphological filter" reflects the fact that they have much in common with morphological filters though being softened.

The two basic soft morphological operations are *soft erosion* and *soft dilation*. Based on these operations, two compound operations, *soft closing* and *soft opening*, are defined.

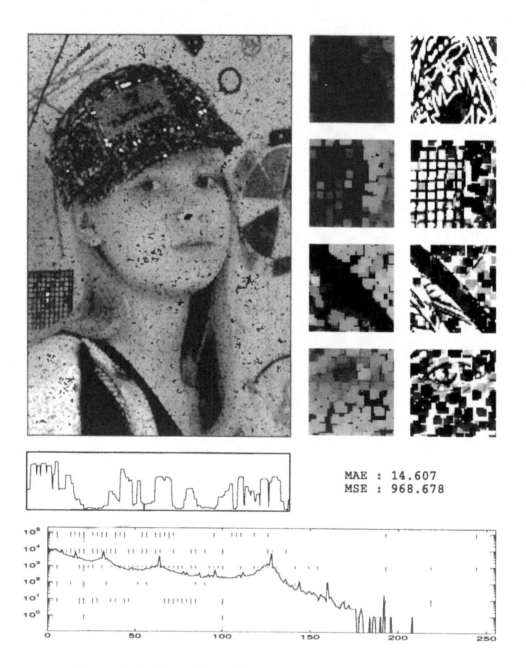

MAE : 14.607
MSE : 968.678

Figure 3.133. "Leena" filtered by the flat opening of window size 3×3.

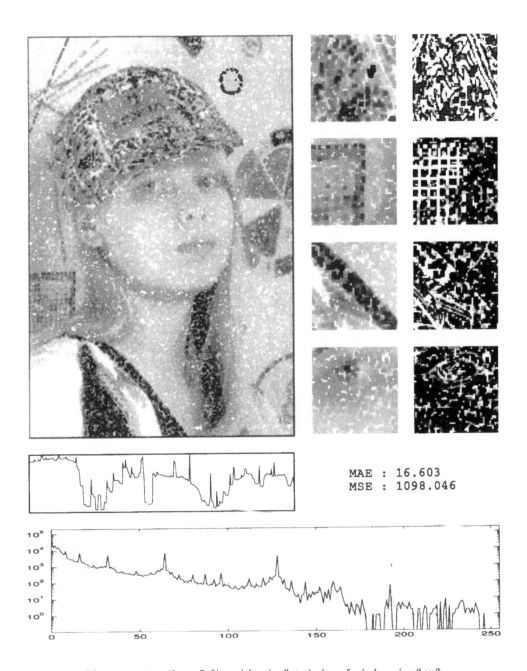

MAE : 16.603
MSE : 1098.046

Figure 3.134. "Leena" filtered by the flat closing of window size 3×3.

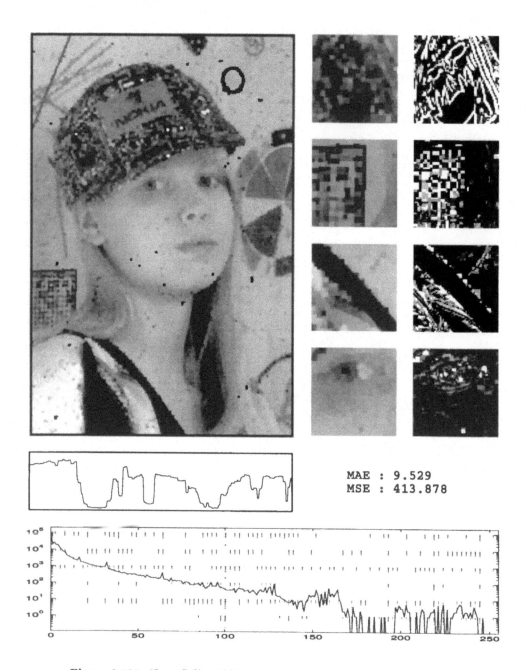

MAE : 9.529
MSE : 413.878

Figure 3.135. "Leena" filtered by the flat open-closing of window size 3 × 3.

Instead of the structuring element of standard flat morphological filters, we now need an extended concept—a *structuring system*. The structuring system $[B, A, r]$ consists of three parameters, finite sets A and B and a natural number r satisfying $1 \leq r \leq |B|$. The set B is called the *structuring set*, A its (hard) *center*, $B \setminus A$ its (soft) *boundary* and r the *order index of its center* or the *repetition parameter*. The structuring set describes the shape and the size of the moving window, $|B| = N$. The hard center describes a special area inside the structuring set, i.e., the samples where "increased" attention is focused.

The soft dilation by the structuring system $[B, A, r]$ is defined by

$$\text{SoftDil}(X_1, X_2, \ldots, X_N; [B, A, r])$$
$$= \text{the } r\text{th largest value of the multiset } \{r \Diamond X_i : i \in A\} \cup \{X_i : i \in (B \setminus A)\}.$$

The soft erosion by the structuring system $[B, A, r]$ is defined by

$$\text{SoftEro}(X_1, X_2, \ldots, X_N; [B, A, r])$$
$$= \text{the } r\text{th smallest value of the multiset } \{r \Diamond X_i : i \in A\} \cup \{X_i : i \in (B \setminus A)\}.$$

If the maximal value inside the moving window is inside the hard center, then it is also the output of the soft dilation. Thus, the hard center acts like the structuring element of the flat dilation. The soft boundary and the order index make it possible for the output to be some other value than the maximum, if the maximum is located in the soft boundary. This gives more robustness to the dilation operation. The soft dilation is biased towards larger values but not to the same extent as the standard dilation. Similarly, the hard center of the soft erosion behaves like the structuring element of the standard flat erosion.

Compound operations are defined in an analogous way to standard morphology: the soft closing by $[B, A, r]$ is defined as the soft dilation by $[B, A, r]$ followed by the soft erosion by $[B^s, A^s, r]$. The soft opening by $[B, A, r]$ is defined as the soft erosion by $[B, A, r]$ followed by the soft dilation by $[B^s, A^s, r]$. Typically, close-openings and open-closings are not defined in soft morphology, since openings and closings can be designed to remove both sided noise.

Example 3.34. Let us choose the structuring system $B = \{-2, -1, 0, 1, 2\}$, $A = \{0\}$, $r = 2$. Now, for the signal $\mathbf{x} = (3, 8, 5, 7, 2, 1, 1, 9, 8, 5)$ we obtain:

original	(3, 8, 5, 7, 2, 1, 1, 9, 8, 5);
soft dilated by $[B, A, r]$	(∗, ∗, 7, 7, 5, 7, 8, 9, ∗, ∗);
soft eroded by $[B, A, r]$	(∗, ∗, 3, 2, 1, 1, 1, 1, ∗, ∗);
soft closed by $[B, A, r]$	(∗, ∗, ∗, ∗, 7, 8, ∗, ∗, ∗, ∗);
soft opened by $[B, A, r]$	(∗, ∗, ∗, ∗, 1, 1, ∗, ∗, ∗, ∗).

Clearly, soft erosion and dilation belong to the class of weighted order statistic filters and thus also to the class of stack filters. So, the algorithm for, e.g., the soft closing can be obtained simply by combining Algorithms 3.10 and 3.27, and they are omitted from this book. Definitions for gray-level structuring systems and resulting soft morphological filters can be obtained in a straightforward way [90].

Some sort of analogy between soft opening and standard opening can be given by considering the rolling ball transformation. In standard opening the ball is hard, staying strictly below the original signal which it is tracing, whereas in soft opening the ball has a harder part that tends to stay below the original signal and a softer boundary which is more "elastic".

Another soft substitute for the classical min/max operators can be obtained by using L_p mean filters. For $p > 0$ we obtain a soft version of erosion by using a L_{-p} mean filter and a soft version of dilation by using a L_p mean filter [85]. These filters have been considered in Section 3.14.

3.18.2 Impulse and Step Response

The soft closing and opening remove impulses if $1 < r \leq |B \setminus A|$ (Exercise). Furthermore, they preserve step edges (Exercise). Thus, soft morphological filters behave better in the presence of impulsive noise than standard morphological filters do.

3.18.3 Filtering Examples

In the one-dimensional example $B = \{-2, -1, 0, 1, 2\}$, $A = \{0\}$, and $r = 3$. In the two-dimensional examples B is the 3×3 square with the origin $(0, 0)$ in the center of it, $A = \{(0, 0)\}$, and $r = 5$.

It is seen that the soft opening can remove both sided impulses, in contrast with the standard opening, and that soft morphological operations can be designed to have good detail preservation properties.

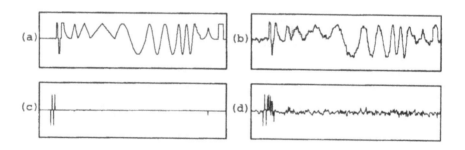

Figure 3.136. One-dimensional signals filtered by the flat soft opening. Parameters are explained in the text.

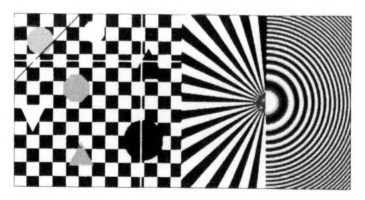

Figure 3.137. "Geometrical" filtered by the flat soft opening. Parameters are explained in the text.

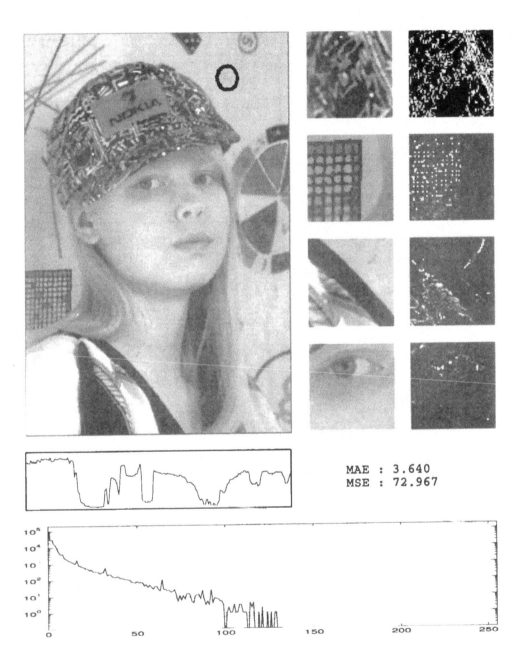

MAE : 3.640
MSE : 72.967

Figure 3.138. "Leena" filtered by the flat soft opening. Parameters are explained in the text.

3.19   Polynomial Filters

3.19.1   Principles and Properties

It is well known that the output of the generalized mean filter in the one-dimensional case can be obtained by using convolution of the input signal $x(n)$ and the unit impulse response $h(n)$, that is,

$$y(n) = \sum_{i=-\infty}^{\infty} h(i)x(n-i).$$

The unit impulse response characterizes the system completely. The unit impulse response also represents the memory of the system and $h(T)$ represents the contribution to the overall output of the input value T time units before the present value.

Multidimensional convolutions can be used to introduce nonlinearities. For example,

$$\sum_{i_1=-\infty}^{\infty} \sum_{i_2=-\infty}^{\infty} h(i_1,i_2)x(n-i_1)x(n-i_2)$$

contains nonlinearities like $h(0,0)x(n)^2$.

A class of nonlinear discrete time-invariant systems with memory is described by means of the *discrete Volterra series*:

$$y(n) = h_0 + \sum_{j=1}^{\infty} \bar{h}_j[x(n)], \tag{3.27}$$

where $x(n)$ is the input, $y(n)$ is the output, h_0 is an offset term, and $\bar{h}_j[x(n)]$ is defined by

$$\bar{h}_j[x(n)] = \sum_{i_1=-\infty}^{\infty} \sum_{i_2=-\infty}^{\infty} \cdots \sum_{i_j=-\infty}^{\infty} h_j(i_1,i_2,\ldots,i_j)x(n-i_1)x(n-i_2)\cdots x(n-i_j). \tag{3.28}$$

Here, $h_1(i_1)$ is the usual linear impulse response and $h_j(i_1,i_2,\ldots,i_j)$ can be considered as a *generalized jth-order impulse response* characterizing the nonlinearities. Discrete Volterra series are derived from *Volterra series* [108, 109, 110]. The functions $h_j(i_1,i_2,\ldots,i_j)$ are called the *Volterra kernels*.

In filtering, the required memory is made finite by limiting all the summations in (3.28) to $N = 2k+1$ samples inside the filter window, e.g., by replacing the summation limits $-\infty$ and ∞ with $-k$ and k, respectively. In a *Volterra (polynomial) filter of order M* the limit ∞ in (3.27) is replaced with M, i.e.,

$$y(n) = h_0 + \sum_{i_1=-k}^{k} h_1(i_1)x(n-i_1) + \cdots$$

$$+ \sum_{i_1=-k}^{k} \sum_{i_2=-k}^{k} \cdots \sum_{i_M=-k}^{k} h_M(i_1,i_2,\ldots,i_M)x(n-i_1)x(n-i_2)\cdots x(n-i_M).$$

$$\tag{3.29}$$

For the corresponding causal filter (the output signal depends only on present and/or previous values of the input), the summations in (3.29) will go from 0 to $N-1$. Article [97] provides a thorough treatise of Volterra filtering and a comprehensive reference list.

The *quadratic filter* is given by the quadratic term ($M = 2$ in (3.29)):

$$y(n) = \sum_{i_1=-k}^{k} \sum_{i_2=-k}^{k} h_2(i_1, i_2) x(n - i_1) x(n - i_2).$$

It is easily seen that in quadratic filtering all possible pairs of input values inside the moving window are selected, and their products are weighted and then summed to give the output. If $i \neq j$, the input pair $x(n - i)$ and $x(n - j)$ is selected twice, for the indices $i_1 = i, i_2 = j$ and $i_1 = j, i_2 = i$. Therefore, typically we assume that $h_2(i, j) = h_2(j, i)$. The coefficients

- $h_2(i, i), -k \leq i \leq k$ give the weights for the samples at distance 0 from each other;

- $h_2(i, i + 1), -k \leq i \leq k - 1$ give the weights for the samples at distance 1 from each other;

  $\vdots$

- $h_2(-k, k)$ give the weights for the samples at distance $N - 1 = 2k$ from each other.

In a similar way we can define the input-output relation for higher dimensional signals. By denoting, as we have usually done, the inputs inside the moving window by $X_1, X_2, \ldots, X_N$ we obtain, e.g., the quadratic filter:

$$\text{Quad}(X_1, X_2, \ldots, X_N; \mathbf{H}) = \sum_{i=1}^{N} \sum_{j=1}^{N} h_{i,j} X_i X_j.$$

This can be written in a compact matrix product form:

$$\text{Quad}(X_1, X_2, \ldots, X_N; \mathbf{H}) = \mathbf{x}\mathbf{H}\mathbf{x}^T$$

$$= (X_1, X_2, \ldots, X_N) \begin{pmatrix} h_{1,1} & h_{1,2} & \cdots & h_{1,N} \\ h_{2,1} & h_{2,2} & \cdots & h_{2,N} \\ \vdots & \vdots & \ddots & \vdots \\ h_{N,1} & h_{N,2} & \cdots & h_{N,N} \end{pmatrix} \begin{pmatrix} X_1 \\ X_2 \\ \vdots \\ X_N \end{pmatrix},$$

where $\mathbf{H}$ is an $N \times N$ matrix (typically symmetric, i.e., $\mathbf{H} = \mathbf{H}^T$). This matrix $\mathbf{H}$ is called the *quadratic kernel* of the filter.

Some special forms of the quadratic kernel are mentioned in Reference [87]:

1. $\mathbf{H}$ is a diagonal matrix; the output is obtained by linear filtering of the squares of the input;

2. $\mathbf{H}$ is dyadic, i.e., $h_{i,j} = h_i h_j$; the output is the square of the output of a linear filter with an impulse response h_i;

3. $\mathbf{H}$ is positive definite; the output is always positive;

4. $\mathbf{H}$ is Toeplitz, i.e., $h_{i,j} = h_{|i-j|}$.

Another useful filter is obtained by combining the FIR-term and the quadratic term:

$$\text{Vol}(X_1, X_2, \ldots, X_N; h, \mathbf{H}) = \sum_{i=1}^{N} h_i X_i + \sum_{i=1}^{N} \sum_{j=1}^{N} h_{i,j} X_i X_j.$$

Polynomial filter

Inputs: *NumberOfRows* × *NumberOfColumns* image
        Moving window $W, |W| = N = 2k + 1$
        Impulse response $\mathbf{h} = (h_1, h_2, \ldots, h_N)$
        2nd-order impulse response $\mathbf{H} = [h_{i,j}]_{N \times N}$
Output: *NumberOfRows* × *NumberOfColumns* image

```
for i = 1 to NumberOfRows
    for j = 1 to NumberOfColumns
        place the window W at (i, j)
        store the image values inside W in x = (X₁, X₂, ..., Xₙ)
        let Sum = 0
        for m = 1 to N
            let Sum = Sum + hₘXₘ
        end
        for m = 1 to N
            for n = 1 to N
                let Sum = Sum + hₘ,ₙXₘXₙ
            end
        end
        let Output(i, j) = Sum
    end
end
```

Algorithm 3.28. Algorithm for the polynomial filter (FIR-term + quadratic term).

For this filter an algorithm is given in Algorithm 3.28. The problem with quadratic and higher order kernels is that they easily add extra gain to the output signal. Thus, they must be used with care.

Example 3.35. Let the 5-point input be $\mathbf{x} = (2, 7, 9, 20, 5)$, the unit impulse response be $\mathbf{h} = (1/5, 1/5, 1/5, 1/5, 1/5)$, and

$$
\mathbf{H} = \frac{1}{25}
\begin{pmatrix}
1/80 & 1/80 & 1/20 & 1/80 & 1/80 \\
1/80 & 1/80 & 1/10 & 1/80 & 1/80 \\
1/20 & 1/10 & 1/5 & 1/10 & 1/20 \\
1/80 & 1/80 & 1/10 & 1/80 & 1/80 \\
1/80 & 1/80 & 1/20 & 1/80 & 1/80
\end{pmatrix}.
$$

The output of the polynomial filter is now

$$
\frac{2 + 7 + 9 + 20 + 5}{5} + (2, 7, 9, 20, 5)\mathbf{H}
\begin{pmatrix}
2 \\
7 \\
9 \\
20 \\
5
\end{pmatrix}
= 8.6 + 3.422 = 12.022.
$$

In order to preserve homogeneous regions with an arbitrary gray-level value a, we must require the FIR-term to satisfy

$$\sum_{i=1}^{N} h_i = 1$$

and the quadratic term to satisfy

$$\sum_{i=1}^{N} \sum_{j=1}^{N} h_{i,j} = 0.$$

The proofs are left as exercises.

The pure quadratic filter cannot preserve arbitrary-valued homogeneous regions because the output equals

$$a^2 \sum_{i=1}^{N} \sum_{j=1}^{N} h_{i,j}$$

and $\sum_{i=1}^{N} \sum_{j=1}^{N} h_{i,j}$ cannot be $1/a$ for all a simultaneously. This could be compensated by dividing the result by $\sum_{i=1}^{N} \sum_{j=1}^{N} h_{i,j}$ (which is then assumed to be nonzero) and then taking the square root.

3.19.2 Impulse and Step Response

The impulse response of the quadratic term is completely defined by the values in the main diagonal of the kernel $\mathbf{H}$. More precisely, the number of nonzero impulse values is equal to the number of nonzero values in the main diagonal.

The quadratic term gives a larger response for steps having the same amplitude but located in brighter regions (0=black, large positive value (255)=white). This feature is in agreement with the well-known Weber's law! In general, polynomial filters also tend to blur steps, as they resemble the averaging operation.

3.19.3 Filtering Examples

We do not show examples now, because in our opinion the polynomial filters do not perform well for impulsive noise removal. This does not mean that they are useless in filtering; on the contrary, they have proven useful for images which, e.g., have been taken in bad illumination conditions and thus are low contrast images [91, 92].

3.20 Data-Dependent Filters

3.20.1 Principles and Properties

Filters whose processing operation is based on some local statistics are often called *data-dependent filters*. The main motivation behind these filters is the fact that signals are seldom stationary, i.e., in different parts of signals the signal characteristics vary. When these variations are noticed by the filter and the filter adapts itself to these variations, improved results can typically be obtained. Even though the word "adapt" is used here, we want to emphasize the difference between adaptive filters and data-dependent filters. Data-dependent filters have zero memory. In other words, they process every window independently. In adaptive filtering the adaptation process has a memory, and the parameter changes also depend on the previous parameters. Anyway, the difference between data-dependent and adaptive filtering is not significant in most cases. Thus, the same filter may be called data-dependent and adaptive.

Local Linear Minimum Mean Square Error Estimator filter

Inputs: *NumberOfRows* × *NumberOfColumns* image
           `Moving window` $W, |W| = N = 2k+1$
           `Noise standard deviation estimate` σ_n
Output: *NumberOfRows* × *NumberOfColumns* image

```
for i = 1 to NumberOfRows
    for j = 1 to NumberOfColumns
        place the window W at (i, j)
        store the image values inside W in x = (X₁, X₂, ..., Xₙ)
        find the median, Med, and the MAD, Mad, of x
```
            `let` $Output(i, j) = \dfrac{Mad^2}{Mad^2 + \sigma_n^2} X^* + (1 - \dfrac{Mad^2}{Mad^2 + \sigma_n^2}) Med$
```
    end
end
```

Algorithm 3.29. Algorithm for the local linear minimum mean square error estimator filter.

Lee [60] has proposed a filtering method based on local statistics. This filtering method has later been shown to be the *local linear minimum mean square error estimator* (LLMMSE) [52] of the original signal S_i, corrupted by additive stationary white noise with zero mean, i.e., $X_i = S_i + N_i$. Let σ_n, σ_x, m be the local estimates of the standard deviations of the noise N_i and the signal X_i, and the local mean inside the moving window, respectively. The output of the local linear minimum mean square error estimator filter is now given by

$$\text{LLMMSE}(X_1, X_2, \ldots, X_N) = (1 - \frac{\sigma_n^2}{\sigma_x^2})X^* + \frac{\sigma_n^2}{\sigma_x^2}m. \qquad (3.30)$$

Under the assumptions of the noise process, $0 \le \sigma_n^2/\sigma_x^2 \le 1$ (see Exercises). When σ_n^2/σ_x^2 varies from 0 to 1, the filter varies from the identity filter to the mean filter. A simple intuitive interpretation of this formula is that for homogeneous areas the standard deviation of the noise is approximately equal to the signal standard deviation, and so the filter is then similar to the mean filter which suppresses Gaussian noise in homogeneous regions well. If, in turn, the signal-to-noise ratio is high, the filter output is approximately the center sample X^*. This happens in signal areas containing details, like edges. However, other detail-like structures, like impulses, will be preserved by the filtering operation, too. The local estimate of the standard deviation of the signal is an indication of our confidence in the local mean estimate and is used to determine the strength of the correction step towards the center sample.

Example 3.36. Let the 5-point input be $\mathbf{x} = (2, 7, 9, 20, 5)$, let us assume that the noise standard deviation is $\sigma_n = 5$, and let us use the sample mean in local estimation of the mean of the signal and (biased) sample variance as the estimate of the variance of the signal. Thus, $m = 8.6$, $\sigma_x^2 = 1/5[(2-8.6)^2 + (7-8.6)^2 + (9-8.6)^2 + (20-8.6)^2 + (5-8.6)^2] = 37.84$, and the output is by (3.30) $9(1 - \sigma_n^2/\sigma_x^2) + 8.6\sigma_n^2/\sigma_x^2 \approx 8.74$.

The performance of the LLMMSE filter depends on the choice of the local estimators for the signal mean and the standard deviations of the signal and noise. The standard deviation of the noise is especially hard to estimate if it changes temporally (spatially). Thus, it is often assumed to be constant.

As discussed in Section 2.2, in image processing the assumption of additive white noise seldom, if ever, holds. Thus, e.g., for impulsive noise filtering the formula (3.30) no longer gives the local linear minimum mean square error estimate. In fact, the correct formula would be [52]

$$\text{LLMMSE} = \frac{\sigma_s^2}{\sigma_s^2 + \sigma_n^2} X^* + (1 - \frac{\sigma_s^2}{\sigma_s^2 + \sigma_n^2})m, \qquad (3.31)$$

where σ_s is the standard deviation of the original signal. We will use this in Algorithm 3.29 where the median is used as the estimate of the mean m and the MAD is used as the estimate of the standard deviation σ_s of the signal. These choices are known to be robust ones. The standard deviation of the noise is assumed to be constant and it is given as a parameter to the filter. Naturally, other location and mean estimates can also be used, cf., e.g., Reference [101].

When a signal is filtered by a mean-type low-pass filter, almost all values in the signal will change. Impulses will change most dramatically, and changes are more profound near edges than in homogeneous signal regions. Thus, the mean filter can be understood as a simple detector of impulses and edges. By giving weights to the samples according to how much they are changed by the low-pass filter, a highly data-dependent nonlinear filter is obtained [41]. For example, for each window, those input samples which are closer to the output of the first filtering operation can be exponentially weighted more. Let the difference of the sample X_i and the result of the low-pass filtering $\hat{X}_i$ at the same position be $|X_i - \hat{X}_i|$. Then this sample has the weight

$$a_i = e^{-\alpha|X_i - \hat{X}_i|},$$

where $\alpha > 0$. The output of the first iteration of the *Average Controlled Local Average filter* (ACLA) is then obtained as a weighted sum of the samples inside the moving window of the filter. This moving window need not be the same window that is used in the calculation of the weights [41]

$$\text{ACLA}(X_1, X_2, \ldots, X_N) = \frac{\sum_{i=1}^N a_i X_i}{\sum_{i=1}^N a_i}.$$

The procedure can be continued by selecting the output of the first iteration to be the reference signal, computing the new weights by comparing the new reference signal to the original signal, and computing the output again using the new weights. This is repeated until the given number of iterations is reached.

Example 3.37. Let the 5-point input be $\mathbf{x} = (2, 7, 9, 20, 5)$. After appending (by repeating the first and the last value of the input) we have the sequence $(2, 2, 7, 9, 20, 5, 5)$. When this appended sequence is filtered by the three-point mean filter, the output sequence is $(3.67, 6, 12, 11.33, 10)$. Using $\alpha = 2$, the weight sequence is found to be $(\exp\{-2|2 - 3.67|\}, \exp\{-2|7 - 6|\}, \exp\{-2|9 - 12|\}, \exp\{-2|20 - 11.33|\}, \exp\{-2|5 - 10|\}) \approx (0.0354, 0.135, 0.00248, 0.000000029, 0.0000454)$. Thus, the second of the samples is considered to be the most trustworthy among the samples, whereas the fourth one is practically discarded as one might expect. The output of the average controlled local average filter is now approximately $(1/0.1729)(0.0354 \cdot 2 + 0.135 \cdot 7 + 0.00248 \cdot 9 + 0.000000029 \cdot 20 + 0.0000454 \cdot 5) \approx 6.0$.

Average Controlled Local Average filter

Inputs: *NumberOfRows* × *NumberOfColumns* image *Original*
      Moving window $W, |W| = N = 2k + 1$
      Control parameter $\alpha > 0$
      Constant natural number *NumberOfIterations*
Output: *NumberOfRows* × *NumberOfColumns* image *Reference*

```
mean filter the image using window W, store the result in Reference image
for k = 1 to NumberOfIterations
    for i = 1 to NumberOfRows
        for j = 1 to NumberOfColumns
            let Weight(i, j) = exp{−α|Original(i, j) − Reference(i, j)|}
        end
    end
    for i = 1 to NumberOfRows
        for j = 1 to NumberOfColumns
            place the window W at (i, j)
            store the values of Original inside W in x = (X₁, X₂, ..., Xₙ)
            store the values of Weight inside W in a = (a₁, a₂, ..., aₙ)
            let Reference(i, j) = (∑ᴺᵢ₌₁ aᵢXᵢ)/(∑ᴺᵢ₌₁ aᵢ)
        end
    end
end
```

Algorithm 3.30. Algorithm for the average controlled local average filter with *NumberOfIterations* iterations.

If the (weighted) mean is replaced with the (weighted) median, *Median Controlled Local Median filters* (MCLM) are obtained. Its algorithm can be easily modified from the algorithm of the average controlled local average filter, Algorithm 3.30. One needs only to change the first *Reference* image to be formed using the median filter, and to change the next *Reference* image calculation to be done by using the weighted median filter with weights a_i.

Naturally, weight functions other than the exponential can be used in a way similar to that in Section 3.11 [41]. This gives more freedom for the filter designer. Furthermore, one can also completely reject potential outliers by letting the weights be zero when the difference between the filtered signal and the original signal exceeds a certain level.

The so called *Recursive Approaching Signal filter* (RASF) is related to the average controlled local average filter. Again, the idea is to find the weights by comparing the reference signal and the original signal, but in the recursive approaching signal filter every sample in a window position is compared with a single reference point, the value of the reference signal at X^* [100]. This is illustrated in Figure 3.139. In the article in Reference [100] the output/new reference image was calculated by the weighted mean operation, but naturally it can be replaced with the weighted median operation. The algorithms of these filters are also easily obtained by modifying Algorithm 3.30.

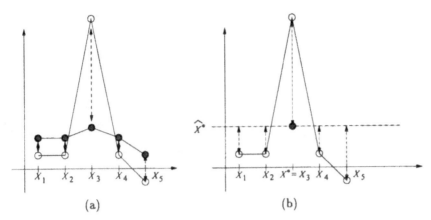

Figure 3.139. The weights for filtering are determined according to the differences between the original signal values (the open circles combined by straight lines) and (a) the reference values (the black bullets) in the average controlled local average filter, and (b) the reference value (the black bullet) in the recursive approaching signal filter.

3.20.2   Impulse and Step Response

As the output of the local linear minimum mean square error estimator filter is a linear combination of the estimate m of the mean and the central pixel X^* in the window of the form $(1 - x)X^* + xm$, the impulse and the step response depend on the mean estimate. If the mean estimate for the window positions containing the impulse is nonzero, the impulse will be spread. Note that the mean estimate will typically be the same wherever the impulse is inside the window, as all the sample values will be the same then. Also the extent to which steps will be blurred depends on the estimates.

The average controlled local average filters and the recursive approaching signal filters give the smallest weight for the impulse. However, for many weight functions, including the exponential one, this weight is nonzero. Thus, while the impulse has an effect on the output the magnitude of the impulse is reduced and the impulse is spread. The median controlled local median filters and the recursive approaching signal filter based on the weighted median operation can have an ideal impulse response. The same holds for steps as well: for most weight functions the median controlled local median filters and the recursive approaching signal filter based on the weighted median operation will preserve steps, whereas the average controlled local average filters and the recursive approaching signal filters based on the weighted mean operation cause moderate blurring.

3.20.3   Filtering Examples

In the filtering examples for the local linear minimum mean square error estimator filter the definition (3.31) is used, the MAD is used as the local estimator of the signal standard deviation, and the median is used to estimate the local mean. The noise variance is fixed and it equals 500 in the one-dimensional examples and the "Leena" example, and it equals 5 in the "Geometrical" example.

The value $\alpha = 0.028$ is used in the examples of the average controlled local average filter and the recursive approaching signal filter based on the weighted mean operation.

These examples show that the local linear minimum mean square error estimator filter is able to remove noise and to keep the signal relatively sharp. It is also seen that those impulses that are not removed by the filter are located just by an edge—area of high signal standard deviation, where more weight is given to the sample X^* than to the mean estimate.

Also the average controlled local average filter and the recursive approaching signal filter based on the weighted mean operation are able to remove impulses, but with this exponential weight function they both cause moderate blurring.

3.21   Decision-Based Filters

3.21.1   Principles and Properties

A useful filtering idea is to include some decision process in filtering. This process decides at every signal position whether or not this position is of a special kind.

Impulse rejecting filters are based on a very simple but effective idea: In every position, first let us decide whether or not we have found an impulse (outlier); if we have found one, let us filter it by some nonlinear filter, otherwise let us leave it unaltered. As we will not filter every pixel, unnecessary distortion can be avoided. However, what is typically lost is the capability of removing Gaussian noise. This idea has been reinvented again and again, but with different impulse detectors. As far as we know, Reference [69] is the first paper describing this idea. We do not try to rank the detectors or even try to list them. A book-length account of the subject of outlier detection has been given by Barnett and Lewis [12].

Perhaps the simplest test is based on the statistic

$$D = \frac{|X^* - X_{(k+1)}|}{s},$$

where X^* is the sample we are making decision of, $X_{(k+1)}$ is the median inside the moving window, and s is some scale estimate. The simplest scale estimate is the sample variance, but it is a very nonrobust estimate. The median of the absolute deviations from the median, MAD,

$$s = 1.483 \text{MED}\{|X_i - \text{MED}\{X_i\}|\}$$

is the most robust estimate of dispersion [30] and can also be used here. The impulse detecting filter with threshold τ is now defined by

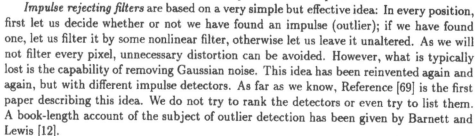

$$\text{ID}(X_1, X_2, \ldots, X_N) = \begin{cases} X^*, & \text{if } D < \tau, \\ X_{(k+1)}, & \text{otherwise.} \end{cases} \qquad (3.32)$$

If the MAD is used one should be aware of one fact. If more than k of the samples inside the window of size $N = 2k + 1$ have the same value the MAD equals zero. This division by zero must be handled somehow. If we decide that this indicates that D is large it means that the pixel under consideration will be filtered. This will lead, e.g., to removal of thin lines. On the other hand, if we decide that in the case of MAD$= 0$ it means that $D < \tau$, then we do not filter the pixel, and impulses in suitably homogeneous regions will not be removed. Thus, in this sense the nonrobust variance may be better or one might use the *t-quantile range* scale estimator

$$R_t = X_{(N+1-t)} - X_{(t)},$$

where $1 \leq t < k$. This estimator may be robust against outliers and against homogeneous areas at the same time. See Algorithm 3.31.

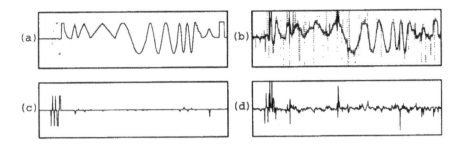

Figure 3.140. One-dimensional signals filtered by the local linear minimum mean square error estimator filter of window length 11, the MAD as the local estimator of the signal standard deviation, the median as the local estimator of the signal local mean, $\sigma_n = 500$.

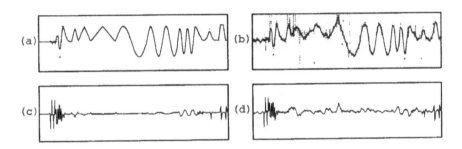

Figure 3.141. One-dimensional signals filtered by the average controlled local average filter of window length 11, $\alpha = 0.028$.

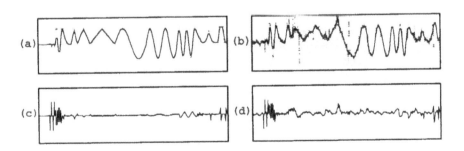

Figure 3.142. One-dimensional signals filtered by the recursive approaching signal filter of window length 11, $\alpha = 0.028$.

Figure 3.143. "Geometrical" filtered by the local linear minimum mean square error estimator filter of window size 5×5, the MAD as the local estimator of the signal standard deviation, the median as the local estimator of the signal local mean, $\sigma_n = 5$.

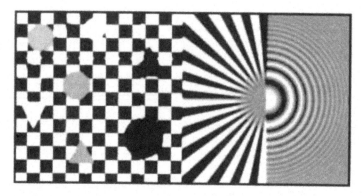

Figure 3.144. "Geometrical" filtered by the average controlled local average filter of window size 5×5, $\alpha = 0.028$.

Figure 3.145. "Geometrical" filtered by the recursive approaching signal filter of window size 5×5, $\alpha = 0.028$.

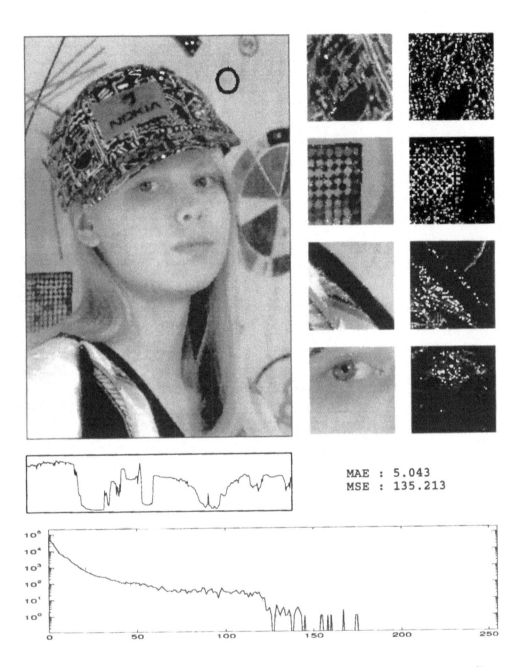

MAE : 5.043
MSE : 135.213

Figure 3.146. "Leena" filtered by the local linear minimum mean square error estimator filter of window size 5 × 5, the MAD as the local estimator of the signal standard deviation, the median as the local estimator of the signal local mean, $\sigma_n = 500$.

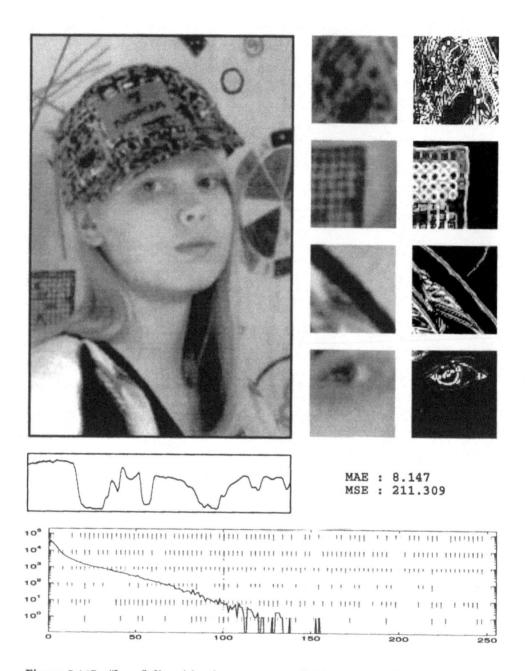

MAE : 8.147
MSE : 211.309

Figure 3.147. "Leena" filtered by the average controlled local average filter of window size 5×5, $\alpha = 0.028$.

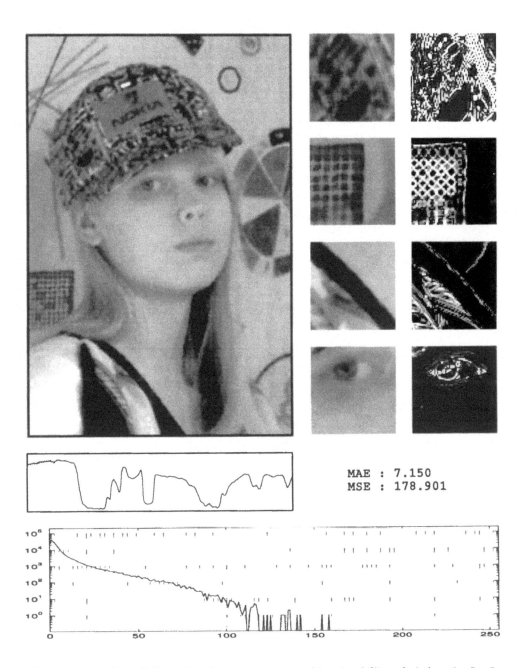

MAE : 7.150
MSE : 178.901

Figure 3.148. "Leena" filtered by the recursive approaching signal filter of window size 5 × 5, $\alpha = 0.028$.

Impulse detecting filter

Inputs: *NumberOfRows* × *NumberOfColumns* image
         Moving window $W, |W| = N = 2k + 1$
         Threshold τ
         Quantile range parameter $t, 1 \leq t < k$
Output: *NumberOfRows* × *NumberOfColumns* image

```
for i = 1 to NumberOfRows
    for j = 1 to NumberOfColumns
        place the window W at (i, j)
        store the image values inside W in x = (X₁, X₂, ..., Xₙ)
        sort x, store the result in y = (X₍₁₎, X₍₂₎, ..., X₍ₙ₎)
        if |X* - X₍ₖ₊₁₎|/(X₍ₙ₊₁₋ₜ₎ - X₍ₜ₎) < τ
            let Output(i, j) = X*
        else
            let Output(i, j) = X₍ₖ₊₁₎
    end
end
```

Algorithm 3.31. Algorithm for an impulse detecting filter.

Another approach is obtained if, after rejecting impulses, the signal is filtered based on the remaining samples. Basically, in this kind of filter, the filtering operation can be any earlier filter. This method is clearly connected with the modified trimmed mean filter discussed in Section 3.2. In fact, the modified trimmed mean filter can be understood as being an impulse rejecting filter, since all the values whose distance from the median is larger than q are considered to be outliers and are rejected from the averaging process.

Naturally, other signal patterns than impulses, like edges, can be detected, and then the obtained information can be used in filtering. Lev *et al.* [67] have applied a template matching technique to detecting edges and lines and then filtered the image by using weighted averages with weights corresponding to the particular pattern detected.

3.21.2   Impulse and Step Response

If the impulse detector works properly, it will signal the presence of the impulse only when we are filtering it and therefore the impulse will not be spread. If the filter used after detection is able to remove impulses, which is typically the case, the impulse will be removed. In the step edge case the impulse detector should not react at all, and thus, the steps will be preserved.

3.21.3   Filtering Examples

The threshold used in the examples is equal to $\tau = 1$. As seen in the examples, most of the impulses are found by the filter, and the details are preserved in an excellent manner. Furthermore, the Gaussian noise in the one-dimensional signal stays unaltered.

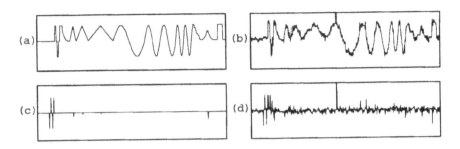

Figure 3.149. One-dimensional signals filtered by the impulse detecting filter (3.32) of window length 11, threshold $\tau = 1$, and quantile-range parameter $t = 4$.

Figure 3.150. "Geometrical" filtered by the impulse detecting filter (3.32) of window size 5×5, threshold $\tau = 1$, and quantile-range parameter $t = 5$.

3.22 Iterative, Cascaded, and Recursive Filters

3.22.1 Principles and Properties

In some of the filters already considered, the filtering process was iterated until the number of iterations was reached. This idea can naturally be used for every filter by repeating the filtering operation until satisfactory results are achieved. In general, it is not easy to give rules for when to stop filtering. The iterative methods can be more effective than the corresponding filter. They can, e.g., improve the noise removal capability of the filter as the signal is being filtered several times instead of one, or they can enhance the edges further. We illustrate this by an example of a ramp edge which is, by iterating the comparison and selection filter, enhanced to a step edge. Some of the iteration results are shown in Figure 3.152. The computational savings achieved by iteration compared with one filter with large window size can also be significant.

Naturally, different filters can also be cascaded, i.e., after processing the signal by the filter A_1, process the resulting signal by the filter A_2 and so on. A natural idea to obtain a filter which removes jointly impulses and Gaussian noise is to first remove the impulses using some filter suited to this task and then filter the resulting image by some filter with good Gaussian noise removal capability. Typically, the performance of the latter one deteriorates severely in the presence of impulses but this is avoided by the first filter.

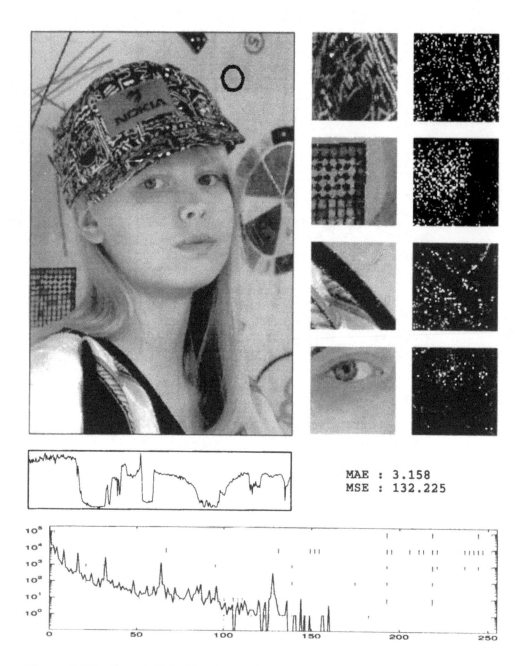

MAE : 3.158
MSE : 132.225

Figure 3.151. "Leena" filtered by the impulse detecting filter (3.32) of window size 5×5, threshold $\tau = 1$, and quantile-range parameter $t = 5$.

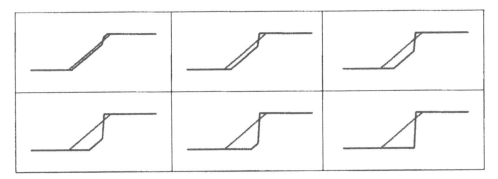

Figure 3.152. A ramp edge enhanced by iterative use of comparison and selection filtering of length 11 and $j = 2$. The dotted line shows the original ramp and the solid line shows the result after filtering. The first row corresponds to the number of iterations, 1,2,4 from left to right, and the second row corresponds to the number of iterations, 6,8,10 from left to right.

The drawbacks of these schemes include the increased danger of losing important details by multiple filtering operations and the increased computational complexity.

To further improve the noise removal capability of a filter, another intuitive modification is to use the already computed samples for the calculation of the output. By this we should have more clean (noise-free) samples inside the window as some of them have already been cleansed by filtering. In this way *recursive filters* are obtained. For example, the recursive median is defined as

$$\text{RecMed}(X_1, X_2, \ldots, X_N) = \text{MED}\{Y_1, Y_2, \ldots, Y_k, X_{k+1}, \ldots, X_N\}, \qquad \text{}$$

where $Y_1, Y_2, \ldots, Y_l$ are the already computed outputs. The recursive filter counterpart for every filter can be obtained in the same way. Especially in the two-dimensional case the properties of the recursive filters will depend on the scanning order of the image.

The cost of the improved noise removal is the loss of detail preservation. The outputs of the recursive filters are highly correlated as some of the outputs are used directly for the computation of others. This will, unfortunately, often lead to increased streaking. Perhaps, the extreme cases could be recursive 1st and Nth ranked-order filters where the output would be disastrous. In contrast to the iterative and cascaded filters, the complexity of recursive filters is not increased.

3.22.2 Impulse and Step Response

Typically, iterative, cascaded, and recursive versions are more immune to impulses than their standard counterparts. Depending on the filters used, the step response will vary.

3.22.3 Filtering Examples

The one-dimensional example of iterative filtering is done using the Hodges-Lehmann D-filter of length 11 iteratively twice. The noise is removed somewhat more efficiently than in the example of the Hodges-Lehmann D-filtering shown in Section 3.12. On the other hand, this iterated use of the Hodges-Lehmann D-filter has resulted in increased rounding of the step edges.

A cascade of the impulse detecting filter (window length 11, threshold $\tau = 1$, and quantile-range parameter $t = 4$) followed by the 3/11-trimmed mean filter (window length 11) is used to illustrate the possible advantages of cascaded filtering. The second filtering operation improved the result of the first operation (see Figure 3.149) quite a lot by attenuating Gaussian noise.

We used the recursive median filter of window length 11 to illustrate recursive filtering. The result of the recursive median filter is smoother than the result of the corresponding nonrecursive median filter (see Figure 1.10). It is known that the result of the one-dimensional recursive median filter will not be changed by iterated filtering [82].

The separable median filter of window size 3×3 is used in the two-dimensional examples of iterative filtering. The results have been essentially improved compared with the one use of the separable median filter of window size 5×5. This is true even though the outputs of these filters are derived using exactly the same 25 samples.

In Figures 3.157 and 3.160 the filter used is the impulse detecting filter used in the previous section followed by the weighted median filter with weights

$$\begin{pmatrix} 1 & 1 & 1 \\ 1 & 5 & 1 \\ 1 & 1 & 1 \end{pmatrix}. \tag{3.33}$$

As can be seen by comparing Figures 3.151 and 3.160 the second filtering operation corrected many of the errors left after the impulse detecting filter.

Finally, we used a recursive median filter of window size 5×5 in the two-dimensional examples of recursive filtering. The results are badly blurred and also the scanning order used in the filtering (start from top left corner, proceed row-wise from left to right) can be seen from the results.

3.23   Some Numerical Measures of Nonlinear Filters

So far, the reader has seen a multitude of nonlinear filters and formed some idea of their behavior. Clearly, every filter has its inherent advantages but has disadvantages as well. Thus, it is impossible to rank different filters. In Reference [89] the behavior of many of the filters studied in this book was summarized in the form of a table containing various figures of merits. Instead of giving a similar table we provide the reader some numerical values measuring noise attenuation and detail preservation capabilities of the studied filters. In Table 3.1 we show the results when the filters in the preceding sections are applied to random 2000×2000 images with independent pixels following three important distributions: Laplacian, Gaussian, and uniform. The mean values of the filtered images are given. From these one can see if the filter is causing shift. The variances of the filtered images are also given illustrating the noise attenuation capability of the filter. The filter parameters are the same as in previous sections when filtering the images "Geometrical" and "Leena". In addition to those filters we have included one extra filter: α-trimmed complementary mean (see Exercises, Section 3.1), which is known to be efficient in uniform distribution. In the table we used the value $\alpha = 1/N$ for the α-trimmed complementary mean filter. We strongly encourage the reader to carefully study the table and compare the effect of the strength of the tails on the noise attenuation capability of the filters.

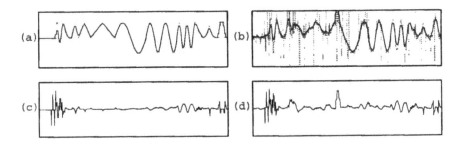

Figure 3.153. One-dimensional signals filtered by the iterative Hodges-Lehmann D-filter of window length 11.

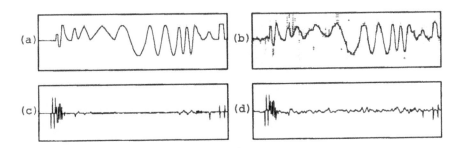

Figure 3.154. One-dimensional signals filtered by the cascade of the impulse detecting filter (window length 11, threshold $\tau = 1$, and quantile-range parameter $t = 4$) followed by the 3/11-trimmed mean filter (window length 11).

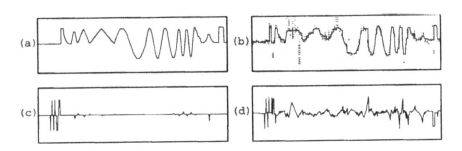

Figure 3.155. One-dimensional signals filtered by the recursive median filter of window length 11.

Figure 3.156. "Geometrical" filtered by the iterative separable median filter of window size 3×3.

Figure 3.157. "Geometrical" filtered by the cascade of the impulse detecting filter (of window size 5×5, threshold $\tau = 1$, and quantile-range parameter $t = 5$) followed by the weighted median filter of window size 3×3 with weights (3.33).

Figure 3.158. "Geometrical" filtered by the recursive median filter of window size 5×5.

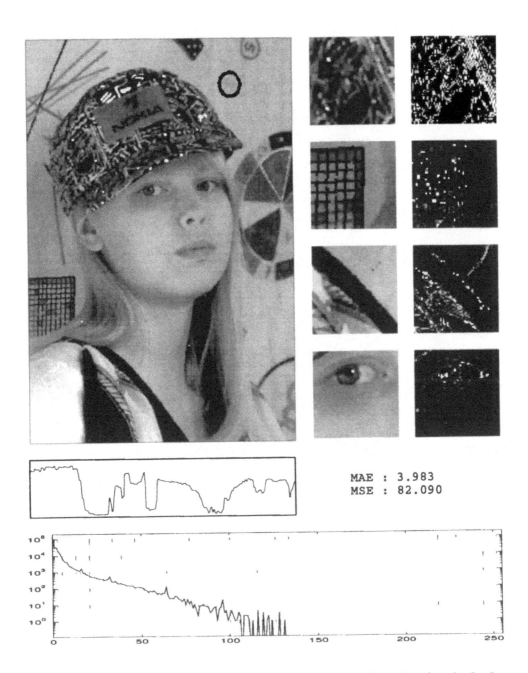

MAE : 3.983
MSE : 82.090

Figure 3.159. "Leena" filtered by the iterative separable median filter of window size 5 × 5.

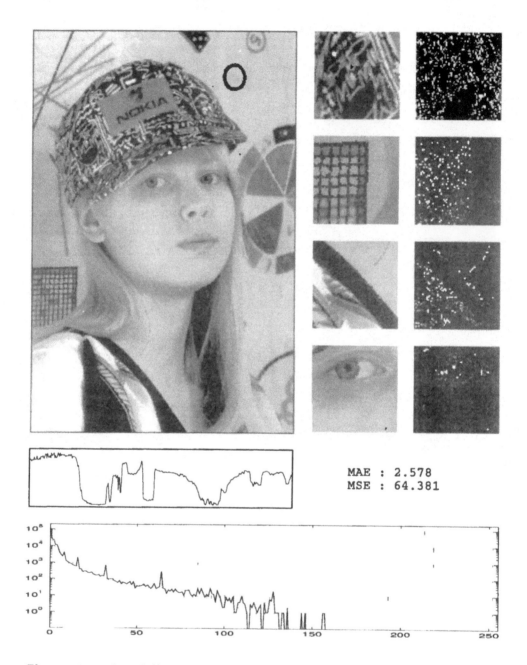

MAE : 2.578
MSE : 64.381

Figure 3.160. "Leena" filtered by the cascade of the impulse detecting filter (of window size 5×5, threshold $\tau = 1$, and quantile-range parameter $t = 5$) followed by the weighted median filter of window size 3×3 with weights (3.33).

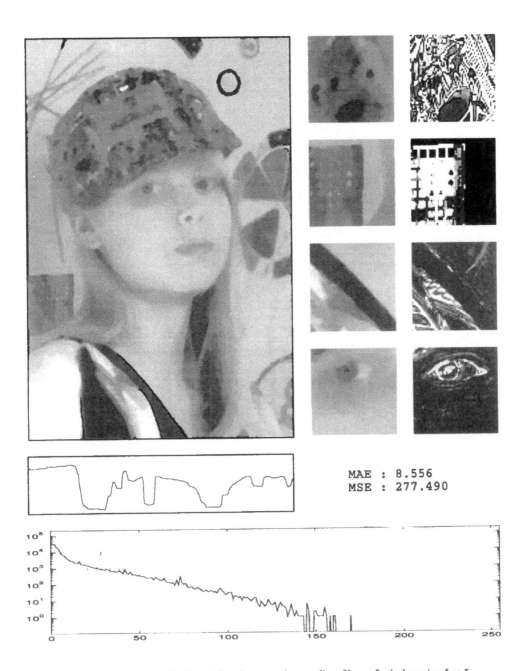

MAE : 8.556
MSE : 277.490

Figure 3.161. "Leena" filtered by the recursive median filter of window size 5 × 5.

Number	Filter	Laplacian		Gaussian		Uniform	
		mean	var	mean	var	mean	var
0	no filter	124.5	99.9	124.5	99.1	124.5	102.2
1	mean	124.0	4.1	124.0	4.0	124.5	4.2
2	median	124.5	2.8	124.5	6.2	125.0	11.4
3	0.4-trimmed mean	124.1	2.6	124.1	5.6	124.1	10.1
4	mod. trimmed mean	124.0	3.8	124.0	4.1	124.0	4.2
5	double window MTM	124.0	3.7	124.0	4.3	124.0	4.2
6	K-nearest neighbor	123.9	34.7	123.9	51.8	123.9	75.5
7	L-filter	124.0	2.6	124.0	5.3	124.0	9.3
8	C-filter	124.0	4.5	124.0	9.0	124.0	15.6
9	weighted median	124.5	6.0	124.5	12.2	124.5	21.3
10	center weighted median	124.5	19.8	124.5	34.0	124.5	54.3
11	ranked-order $R = 7$	119.7	5.8	118.2	7.2	116.4	9.0
12	ranked-order $R = 19$	129.3	5.8	130.8	7.2	132.6	8.9
13	weighted order stat	119.7	12.5	118.3	14.2	116.8	16.4
14	separable median	124.5	3.8	124.5	8.2	124.5	14.9
15	MSM5	124.5	29.5	124.5	44.5	124.5	64.8
16	MSM6	124.5	9.5	124.5	17.6	125.0	29.7
17	max/median	128.2	12.8	129.3	17.1	131.1	23.2
18	1LH+	124.3	12.1	124.3	16.2	124.3	19.7
19	R1LH+	124.3	12.1	124.3	16.2	124.3	19.8
20	2LH+	124.3	14.3	124.3	19.7	124.3	25.0
21	Hachimura-Kuwahara	124.1	11.3	124.1	18.1	124.1	28.7
22	comparison & selection	124.5	4.7	124.5	12.7	125.0	26.1
23	RCRS	125.0	25.8	125.2	39.5	125.4	57.7
24	LUM smoother	124.5	26.5	124.5	42.7	124.5	65.1
25	LUM sharpener	124.3	112.8	124.3	120.4	124.4	133.9
26	LUM filter	124.5	26.7	124.5	43.1	124.5	65.8
27	standard type M	124.0	3.0	124.0	4.5	124.0	7.2
28	Tukey's biweight M	124.0	3.1	124.0	4.4	124.0	5.0
29	gamma 0.7	124.5	3.6	124.5	9.3	124.5	19.3
30	gamma 1.2	124.5	2.7	124.5	5.5	124.5	9.1
31	Wilcoxon	124.3	3.0	124.3	4.3	124.8	4.9
32	Hodges-Lehmann	124.7	3.2	124.3	4.3	124.3	4.8
33	WMMR^m	124.0	5.3	124.0	16.8	124.0	56.1
34	geometric mean	123.6	4.2	123.6	4.1	124.1	4.2
35	harmonic mean	123.2	4.5	123.2	4.1	123.7	4.2
36	L_p mean	122.8	6.1	122.9	4.4	122.9	4.2
37	contraharm. mean	121.5	8.4	121.7	4.7	121.6	4.2
38	cascade of L_p & L_{-p} mean	123.4	11.3	123.4	11.1	123.4	11.1
39	stack	124.5	8.8	124.5	16.5	124.5	27.8
40	flat opening	116.6	32.3	115.4	23.5	114.2	16.8
41	flat closing	132.4	32.2	133.6	23.6	134.8	16.7
42	open-closing	118.4	10.6	116.8	11.5	115.1	12.4
43	flat soft opening	123.6	8.8	123.3	16.3	122.9	27.2
44	LLMMSE	124.0	3.2	124.0	6.9	124.0	12.1
45	ACLA	123.9	2.8	123.9	3.6	123.9	4.3
46	RASF	123.9	3.3	123.9	4.2	123.9	4.9
47	impulse detecting	124.5	41.8	124.5	70.7	124.5	99.5
48	iterative separable	124.5	6.7	124.5	13.2	124.5	22.6
49	cascaded	124.5	16.3	124.5	33.5	124.5	59.3
50	recursive median	124.6	0.3	124.6	0.3	124.4	0.4
51	$1/N$-trimmed complementary mean	124.2	40.9	124.3	13.2	124.3	1.0

Table 3.1. Mean values and variances of the filtering results of the experimental 2000×2000 images with Laplacian, Gaussian, and uniformly distributed samples.

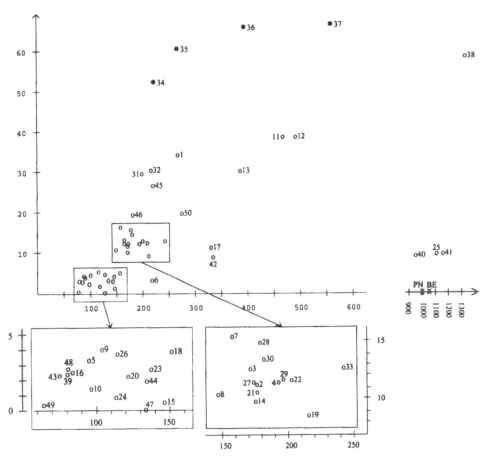

Figure 3.162. Noise attenuation versus detail preservation. The horizontal axis corresponds to the MSE after filtering the noisy "Leena" and the vertical axis corresponds to the MAE after filtering the image "Geometrical". The numbers of the filters are shown in Table 3.1 and their parameters were given in the corresponding sections. BE denotes the noisy image (both sided) and PN denotes the noisy images (only positive errors, used for the filters 34–37).

Measuring visual quality is even more difficult than quantifying noise attenuation. Figure 3.162 illustrates the detail preservation properties versus the bit error attenuation properties of the studied filters. The horizontal axis corresponds to the MSE after filtering the noisy "Leena" (Figure 1.15 or Figure 3.113 for filters 34–37). These numbers have been given in the corresponding figures. Thus, the horizontal axis is used to indicate noise attenuation in a real image; the smaller the value is, the better the noise is removed. The vertical axis corresponds to the MAE after filtering the image "Geometrical". As we did not have any noise in the image "Geometrical", the MAE gives an indicator of detail preservation. This measure is now obtained from a real image. In general it is difficult to derive proper measures of detail preservation, and the one used here is one possibility whose drawback is that it depends on the image used.

3.24   Discussion

Several nonlinear filters have been described in this chapter. The motivations behind these filters were explained, and their principles and basic properties were studied. Some filters result from efforts to find a good compromise between the median and the mean filters. In our opinion most of these compromises are relatively successful. Another group of filters can be categorized as ad-hoc techniques. They are typically characterized by an excellent behavior in a specific limited task and a failure to behave well outside it. Still, we believe that they have their own place and applications among nonlinear filters. A clear drawback of these filters is that it is often not possible to analyze their behavior because of their specialized structure. One important group of filters originates from robust statistics. These filters have many desirable properties, and we believe that in most of the filtering cases the application engineer should consider applying some of these filters. Especially, M-filters and their modifications should be studied more thoroughly as they have great potential. Some of the studied filters, like polynomial filters, perhaps better suit applications other than pure noise removal, like channel equalization or modeling of nonlinear systems.

We have been asked in our lectures which filter is the best. Our answer has been: "It depends on the application", which is clearly not only a typical answer of a professional politician but also the correct answer. If we are asked about our preferences, we would choose tractable filters where their statistical analysis is possible, providing us with the tools for finding the successful filtering operation. Furthermore, the filter class should be relatively large providing enough degrees of freedom to work with. Still, we do not like too general definitions resulting in filter classes containing too different types of filters.

Writing this chapter was not an easy task and we definitely realize that our choices, interpretations, and comments can be questioned. Hopefully, we were able to reveal the wide spectrum of ideas that have been used so far. But this is not the end of the story. We summarize our hopes of the future in the following slogan:

> IT IS IMPOSSIBLE FOR A MAN WHO TAKES A SURVEY OF WHAT IS ALREADY KNOWN, NOT TO SEE WHAT AN IMMENSITY IN EVERY BRANCH OF SCIENCE YET REMAINS TO BE DISCOVERED.
> Thomas Jefferson, 1799

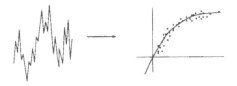

STATISTICAL ANALYSIS AND OPTIMIZATION OF NONLINEAR FILTERS

THERE IS NOTHING MORE PRACTICAL THAN A GOOD THEORY
Unknown scientist

IF YOU HAVE BUILT CASTLES IN THE AIR, YOUR WORK NEED NOT
TO BE LOST, THAT IS WHERE THEY SHOULD BE—NOW PUT A
FOUNDATION UNDER THEM
H. D. Thoreau, 1817–1862

4.1 Methods Based on Order Statistics

In Chapter 2 we saw that the most fruitful way to describe and analyze image processing
methods is to model both the image and the corrupting interference as stochastic processes
or, because we only use digitized images, as random vectors. If we are able to derive the
distribution function of the random variable that represents the output of a filtering
operation (or, more exactly, the joint distribution function of the original underlying
signal and the output) we can characterize the error in a quantitative statistical sense. If
the expression for the distribution is a manageable function of the filter parameters we
could try to find a closed form solution for the optimal filter with respect to a predefined
error criterion. Even if we are only able to numerically compute the output distribution
we can still use numerical methods to find an approximation of the optimal solution as
accurately as necessary. If the filtering operation is so complicated that neither of the

above is possible we can only use heuristical search algorithms to find a good filter. The fundamental problem with the last approach is that even though it usually works fairly well we really have no idea about how good the resulting filter is after all.

The theory of nonlinear filtering is still in its infancy and we know the exact statistical characterizations only for very few filters and even then only for the simplest cases. The simplest but an important case is to find the output distribution of the filter when the input is white noise, i.e., a sequence of independent and identically distributed random variables. It is often also assumed that the noise is Gaussian, but for most nonlinear filters the difficulty of deriving the output distribution is quite independent of the actual type of the noise distribution. Once we have the output distribution we can immediately compute all the relevant characteristics such as the expectation, which reveals whether the filter gives a systematic error or not, and the variance, which gives a clear indication of the noise attenuation power of the filter. If the expectation of the output deviates from the expectation of the input the filter systematically increases or decreases the signal level. In images this means that the output image is lighter or darker than the original image. The variance of the signal is directly linked to the power of the signal. Thus, a filter whose output variance is small for white noise input reduces the energy of noise. In images this means that after filtering the image looks less rough. However, it is important to remember that the visual appearance of the filtered images depends much on factors other than the noise attenuation capability of the filter.

In the following, we derive the output distributions of some of the filters introduced in the previous chapter. The approach is somewhat different. While we earlier aimed to describe each filter in detail and explain their particular strengths and weaknesses, we here strive for general methods that can be used to handle large classes of filters. We believe that this best reveals the similarities and differences of the various filters and gives the student a much deeper understanding also of the particular filtering methods.

We shall concentrate on two general methods. The first is based on determining the joint distribution of the order statistics of the random variables in the filter window. The classic book by S. Wilks [116] is the main reference here. The joint distribution can be fairly easily derived using the Dirichlet distribution that was introduced in Chapter 2. From the joint distribution of the order statistics we can immediately write down the formulas for the output distributions of L-filters including all trimmed means and ranked order filters. The second method is based on representing the filter as a stack filter. The general formulas that exist for the output distribution of stack filters can then be used to derive the output distribution of the particular filter in question. Because many nonlinear filters, e.g., multistage median filters and median hybrid filters, can be interpreted as stack filters operating on more general input signals, the formulas originally developed for stack filters lend themselves for analyzing these filter classes.

In this chapter we are able to perform statistical characterization for only a few of the filter classes that we introduced in Chapter 3, not to mention the large number of different filtering methods that have not been included in this book. The reasons for selecting the particular filter classes to be analyzed is twofold. Firstly, L-filters and stack filters perform quite well in many applications and are, for that reason, widely used. Secondly, together they form quite a large class of filters, and with suitable tricks many filters can be viewed as members of these classes and so also analyzed using these methods. The presentation is adapted mainly from the references that were pointed out in the previous chapter when introducing the particular filter classes. Additional references are shown when necessary.

4.1.1 Joint Distributions of Order Statistics

Let X be a random variable with a continuous distribution $F(t)$. Then the random variable $Y = F(X)$ has the uniform distribution $U(0,1)$ whose density is equal to 1 for $0 < t < 1$ and zero otherwise. Formally, this is evident from

$$P(Y \le t) = P(F(X) \le t) = P(X \le F^{-1}(t)) = F(F^{-1}(t)) = t.$$

Let now $X_1, X_2, \ldots, X_N$ be i.i.d. random variables with the common distribution function $F(t)$ and let $Y_i = F(X_i)$, $i = 1, 2, \ldots, N$. Then the probability element

$$P((Y_1, Y_2, \ldots, Y_N) \in [y_1, y_1 + dy_1] \times [y_2, y_2 + dy_2] \times \cdots \times [y_N, y_N + dy_N])$$

is $1 \cdot dy_1 dy_2 \cdots dy_N$ for $(y_1, y_2, \ldots, y_N) \in [0,1]^N$ (the unit cube) and $0 \cdot dy_1 dy_2 \cdots dy_N$ outside the unit cube. The random vector $(Y_{(1)}, Y_{(2)}, \ldots, Y_{(N)})$, where the components are ordered in the increasing order, is called the *order statistic* of the random vector $(Y_1, Y_2, \ldots, Y_N)$, and we consider the problem of determining its probability element. Because we assume that the distribution is continuous the probability that two or more elements in the sample are equal is zero. So, we may consider only the points in the unit cube whose coordinates are all distinct. Let $(y_1, y_2, \ldots, y_N)$ be such a point P. Because all y_i are different altogether $N!$ distinct points are obtained by permuting the coordinates of P. If a new random variable $(Y_{(1)}, Y_{(2)}, \ldots, Y_{(N)})$ is formed by always ordering the components of $(Y_1, Y_2, \ldots, Y_N)$ its probability element is obtained by adding all permutations of $1 \cdot dy_1 dy_2 \cdots dy_N$ and using the ordering notation. Thus, the probability element of $(Y_{(1)}, Y_{(2)}, \ldots, Y_{(N)})$ is

$$N! dy_{(1)} dy_{(2)} \cdots dy_{(N)} \tag{4.1}$$

inside the region $0 < y_{(1)} < y_{(2)} < \cdots < y_{(N)} < 1$ and $0 \cdot dy_{(1)} dy_{(2)} \cdots dy_{(N)}$ outside it. Thus, the density of $(Y_{(1)}, Y_{(2)}, \ldots, Y_{(N)})$ is

$$f(y_{(1)}, y_{(2)}, \ldots, y_{(N)}) = \begin{cases} N!, & \text{for } 0 < y_{(1)} < y_{(2)} < \cdots < y_{(N)} < 1, \\ 0, & \text{otherwise.} \end{cases} \tag{4.2}$$

But this can be written as

$$f(y_{(1)}, y_{(2)}, \ldots, y_{(N)})$$
$$= \begin{cases} \frac{\Gamma(N+1)}{\Gamma(1) \cdots \Gamma(1)} y(1)^0 \cdots y(N)^0 (1 - y_{(1)} - \cdots - y_{(N)})^0, & \text{for } 0 < y_{(1)} < \cdots < y_{(N)} < 1, \\ 0, & \text{otherwise.} \end{cases}$$
$$\tag{4.3}$$

Thus we arrive at the following result. Let $(X_{(1)}, X_{(2)}, \ldots, X_{(N)})$ be the order statistics from a continuous distribution with the distribution function $F(t)$. Then the random vector $(F(X_{(1)}), F(X_{(2)}), \ldots, F(X_{(N)}))$ has the ordered N-dimensional Dirichlet distribution $D^*(1, \ldots, 1; 1)$. This is, of course, nothing but the uniform distribution in the simplex $\sum_{i=1}^{N} y_{(i)} < 1$ but by the properties of Dirichlet distribution this result gives immediately the distributions of subsets of $\{F(X_{(1)}), F(X_{(2)}), \ldots, F(X_{(N)})\}$.

The random variable $F(X_{(k)})$ has the Beta distribution $B(k, N - k + 1)$ and for the case of s of these random variables the random vector

$$(F(X_{(k_1)}), F(X_{(k_1+k_2)}), \ldots, F(X_{(k_1+k_2+\cdots+k_s)})))$$

has the ordered Dirichlet distribution $D^*(k_1, k_2, \ldots, k_s; N - k_1 - \cdots - k_s + 1)$.

Now that we have the joint probability elements of subsets of order statistics of uniformly distributed random variables $F(X_{(1)}), F(X_{(2)}), \ldots, F(X_{(N)})$ we can immediately form the corresponding probability elements of the order statistics $X_{(1)}, X_{(2)}, \ldots, X_{(N)}$ provided that the density $f(t) = \frac{d}{dt} F(t)$ exists. Directly from (4.1) we see that if $X_{(1)}, X_{(2)}, \ldots, X_{(N)}$ are the order statistics of N i.i.d. random variables with a common distribution $F(t)$ and density $f(t)$ then the joint probability element of $X_{(1)}, X_{(2)}, \ldots, X_{(N)}$ is

$$N! f(x_{(1)}) f(x_{(2)}) \cdots f(x_{(N)}) dx_{(1)} dx_{(2)} \cdots dx_{(N)} \tag{4.4}$$

in the region $-\infty < x_{(1)} < x_{(2)} < \cdots < x_{(N)} < \infty$ and zero outside. The probability elements of subsets of $X_{(1)}, X_{(2)}, \ldots, X_{(N)}$ follow as above. The probability element of $X_{(k)}$ is

$$\frac{\Gamma(N+1)}{\Gamma(k)\Gamma(N-k+1)} F(x_{(k)})^{k-1} (1 - F(x_{(k)}))^{N-k} f(x_{(k)}) dx_{(k)} \tag{4.5}$$

for $-\infty < x_{(k)} < \infty$ and the probability element of $X_{(k_1)}, X_{(k_1+k_2)}, \ldots, X_{(k_1+k_2+\cdots+k_s)}$ is

$$\frac{\Gamma(N+1)}{\Gamma(k_1)\Gamma(k_2)\cdots\Gamma(k_s)\Gamma(N-k_1-k_2-\cdots-k_s+1)} \times$$
$$[F(x_{(k_1)})]^{k_1-1} [F(x_{(k_1+k_2)}) - F(x_{(k_1)})]^{k_2-1} \cdots [1 - F(x_{(k_1+k_2+\cdots+k_s)})]^{N-k_1-k_2-\cdots-k_s} \times \tag{4.6}$$
$$f(x_{(k_1)}) \cdots f(x_{(k_1+k_2+\cdots+k_s)}) dx_{(k_1)} dx_{(k_1+k_2)} \cdots dx_{(k_1+k_2+\cdots+k_s)}$$

in the region $-\infty < x_{(k_1)} < x_{(k_1+k_2)} < \cdots < x_{(k_1+k_2+\cdots+k_s)} < \infty$ and zero outside. These formulas were originally derived by Craig [22] and they contain many special cases that are useful in the study of filters based on order statistics. If we set $k = N$ we obtain the probability element of the largest element of $X_1, X_2, \ldots, X_N$,

$$N F(x_{(N)})^{N-1} f(x_{(N)}) dx_{(N)},$$

if we set $k = 1$ we obtain the probability element of the smallest element of $X_1, X_2, \ldots, X_N$,

$$N (1 - F(x_{(1)}))^{N-1} f(x_{(1)}) dx_{(1)}$$

and if N is odd and we set $k = (N+1)/2$ we obtain the probability element of the median of $X_1, X_2, \ldots, X_N$

$$\frac{(2k+1)!}{k! k!} F(x_{(k+1)})^k (1 - F(x_{(k+1)}))^k f(x_{(k+1)}) dx_{(k+1)}.$$

An important case also follows from the choice $s = 2$, $k_1 = 1$, and $k_2 = N-1$ in (4.6) giving the joint probability element of the smallest and the largest element of $X_1, X_2, \ldots, X_N$, i.e.,

$$N(N-1)[F(x_{(N)}) - F(x_{(1)})]^{N-2} f(x_{(1)}) f(x_{(N)}) dx_{(1)} dx_{(N)}. \tag{4.7}$$

4.1.2   Analysis of L-Filters

L-filters were studied in Chapter 3 and we saw that ranked order filters, including the median filter, are special cases of L-filters. Also both standard and trimmed means could be expressed as L-filters. In the following we derive formulas for computing the output distribution and the moments of an L-filter. From the covariance matrix of order statistics under an input distribution we can then determine the optimal coefficients in the mean square sense.

Suppose that $X_1, X_2, \ldots, X_N$ are i.i.d. random variables with the common distribution function $F(t)$ and density $f(t)$. By (4.2) the joint density of the order statistics $X_{(1)}, X_{(2)}, \ldots, X_{(N)}$ is

$$g(x_{(1)}, x_{(2)}, \ldots, x_{(N)}) = N! f(x_{(1)}) f(x_{(2)}) \cdots f(x_{(N)}) \tag{4.8}$$

in the region $-\infty < x_{(1)} < x_{(2)} < \cdots < x_{(N)} < \infty$ and zero outside. Now, the output of the L-filter with coefficients $a_1, a_2, \ldots, a_N$, is

$$Y = \sum_{i=1}^{N} a_i X_{(i)}, \tag{4.9}$$

and so the distribution function of Y is

$$G(y) = \int \cdots \int_D N! f(x_{(1)}) f(x_{(2)}) \cdots f(x_{(N)}) dx_{(1)} dx_{(2)} \cdots dx_{(N)}, \tag{4.10}$$

where $D = \{(x_{(1)}, x_{(2)}, \ldots, x_{(N)}) : x_{(1)} < x_{(2)} < \cdots < x_{(N)}, \sum_{i=1}^{N} a_i x_{(i)} \leq y\}$.

As the output of the L-filter (4.9) is a function of the random variables $X_{(1)}, X_{(2)}, \ldots, X_{(N)}$ whose joint density is (4.8) we can compute any characteristics, e.g., the expectation and the variance of (4.9) directly from (4.8).

Consider the computation of the moments of (4.9). Now,

$$E\{Y^m\} = E\{(\sum_{i=1}^{N} a_i X_{(i)})^m\}$$

$$= \int \cdots \int_D (\sum_{i=1}^{N} a_i x_{(i)})^m N! f(x_{(1)}) f(x_{(2)}) \cdots f(x_{(N)}) dx_{(1)} dx_{(2)} \cdots dx_{(N)}$$

where $D = \{(x_{(1)}, x_{(2)}, \ldots, x_{(N)}) : -\infty < x_{(1)} < x_{(2)} < \cdots < x_{(N)} < \infty\}$. Making the transformation $y_{(i)} = F(x_{(i)})$, $i = 1, 2, \ldots, N$ yields

$$E\{Y^m\} = \int \cdots \int_F \left(\sum_{i=1}^{N} a_i F^{-1}(y_{(i)}) \right)^m N! dy_{(1)} dy_{(2)} \cdots dy_{(N)}. \tag{4.11}$$

Because for $i = 1, 2, \ldots, N$ the transformation $y_{(i)} = F(x_{(i)})$ maps the interval $(-\infty, \infty)$ to the interval $(0, 1)$ and maintains the order, the region of integration transforms into $F = \{(y_{(1)}, y_{(2)}, \ldots, y_{(N)}) : 0 < y_{(1)} < y_{(2)} < \cdots < y_{(N)} < 1\}$.

In the actual computations it is convenient to tabulate $F^{-1}(y)$ at uniformly spaced points in the interval $[0, 1]$ and, if necessary, use interpolation for intermediate values. An advantage of (4.11) is that the integration region is fixed and only a_i, F, and m need to be specified.

Example 4.1. Let us compute the expectation of the L-filter

$$Y = a_1 X_{(1)} + a_2 X_{(2)} + a_3 X_{(3)},$$

where X_1, X_2, X_3 are independent and uniformly distributed on $[0, 1]$, $a_i > 0$, and $a_1 + a_2 + a_3 = 1$. Then

$$E\{Y\} = \int \int \int_D (a_1 x_{(1)} + a_2 x_{(2)} + a_3 x_{(3)}) 3! \, dx_{(1)} dx_{(2)} dx_{(3)}, \tag{4.12}$$

where $D = \{(x_{(1)}, x_{(2)}, x_{(3)}) : 0 < x_{(1)} < x_{(2)} < x_{(3)} < 1\}$. The transformation

$$\begin{cases} y_1 = x_{(1)}, \\ y_2 = x_{(2)} - x_{(1)}, \\ y_3 = x_{(3)} - x_{(2)} \end{cases}$$

has its Jacobian equal to unity and transforms the integration region D to F (Exercise):

$$\begin{cases} 0 < y_1 < 1, \\ 0 < y_2 < 1 - y_1, \\ 0 < y_3 < 1 - y_1 - y_2. \end{cases}$$

The integral (4.12) can then be written as a simple iterated integral

$$\begin{aligned} E\{Y\} &= \int_0^1 \int_0^{1-y_1} \int_0^{1-y_1-y_2} [(a_1 + a_2 + a_3)y_1 + (a_2 + a_3)y_2 + a_3 y_3]\, 3!\, dy_1\, dy_2\, dy_3 \\ &= \frac{1}{4}(a_1 + 2a_2 + 3a_3). \end{aligned}$$

Remark 4.1. The linearity of the expectation operation implies that (4.12) can in fact be written as

$$E\{Y\} = a_1 E\{X_{(1)}\} + a_2 E\{X_{(2)}\} + a_3 E\{X_{(3)}\}.$$

Because X_1, X_2, X_3 are independent and uniformly distributed on $[0, 1]$, each order statistic $X_{(k)}$, $k = 1, 2, 3$ has the Beta distribution $B(k, 4 - k)$. The expectation of $B(\alpha, \beta)$ is $\frac{\alpha}{\alpha+\beta}$ giving

$$E\{Y\} = a_1 \cdot \frac{1}{4} + a_2 \cdot \frac{2}{4} + a_3 \cdot \frac{3}{4} = \frac{1}{4}(a_1 + 2a_2 + 3a_3).$$

In general it follows from (4.5) that for $X_1, X_2, \ldots, X_N$ with a common distribution function F and density f the density $f_k(x_{(k)})$ of $X_{(k)}$ is

$$f_k(x_{(k)}) = N \binom{N-1}{k-1} F(x_{(k)})^{k-1} \left(1 - F(x_{(k)})\right)^{N-k} f(x_{(k)}), \qquad (4.13)$$

which can also be derived directly (Exercise).

The formula (4.11) can be used to compute the correlation matrix of $X_{(i)}$ when $X_1, X_2, \ldots, X_N$ are i.i.d. random variables. It is the $N \times N$ matrix $\mathbf{R}$, where

$$\mathbf{R} = \left[E\{X_{(i)} X_{(j)}\}\right], \quad i, j = 1, 2, \ldots, N.$$

The high dimension of the region of integration makes (4.11) unsuitable for numerical computations if N is large. For large N it is often better to extract from (4.6) the joint density function of $X_{(i)}$ and $X_{(j)}$:

$$\begin{aligned} & f_{ij}(x_{(i)}, x_{(j)}) \\ &= \frac{\Gamma(N+1)}{\Gamma(i)\Gamma(j-i)\Gamma(N-j+1)} F(x_{(i)})^{i-1} (F(x_{(j)}) - F(x_{(i)}))^{j-i-1} (1 - F(x_{(j)}))^{N-j} f(x_{(i)}) f(x_{(j)}) \\ &= \frac{N!}{(i-1)!(j-i-1)!(N-j)!} F(x_{(i)})^{i-1} (F(x_{(j)}) - F(x_{(i)}))^{j-i-1} (1 - F(x_{(j)}))^{N-j} f(x_{(i)}) f(x_{(j)}) \end{aligned}$$

$$(4.14)$$

for $x_{(i)} < x_{(j)}$ and zero otherwise.

4.1.3 Optimization of *L*-Filters

In the following, we briefly describe how to find the optimal coefficients a_i when we know the second order statistics of the signal and the noise [18, 28]. Consider the signal

$$X_n = S_n + N_n$$

and assume that the samples $X_1, X_2, \ldots, X_N$ are in the window. We wish to find $a_1, a_2, \ldots, a_N$ that minimize the mean square error

$$E\{(Y - S_k)^2\},$$

where

$$Y = a_1 X_{(1)} + a_2 X_{(2)} + \cdots + a_N X_{(N)},$$

and k denotes the reference point (usually the center) of the window (i.e., we are estimating S_k from the data $X_1, X_2, \ldots, X_N$). Now, we can write

$$E\{(Y - S_k)^2\} = \mathbf{a}^T \mathbf{R} \mathbf{a} - 2\mathbf{a}^T \mathbf{r} + E\{S_k^2\},$$

where

$$\mathbf{R} = \left[E\{X_{(i)} X_{(j)}\} \right], \quad i, j = 1, 2 \ldots, N$$

is the correlation matrix of the order statistics $X_{(1)}, X_{(2)}, \ldots, X_{(N)}$, and

$$\mathbf{r} = (E\{S_k X_{(1)}\}, E\{S_k X_{(2)}\}, \ldots, E\{S_k X_{(N)}\})^T$$

is the cross correlation between S_k and the data. If we require *location invariance*, i.e., $X_n' = X_n + c$ implies $E\{Y'\} = E\{Y\} + c$ then $\mathbf{a} = (a_1, a_2, \ldots, a_N)^T$ must satisfy

$$\mathbf{a}^T \mathbf{1} = 1, \tag{4.15}$$

where $\mathbf{1} = (1, \ldots, 1)^T$. If we require that Y is *unbiased*, i.e., $E\{Y\} = E\{S_k\}$, then $\mathbf{a} = (a_1, a_2, \ldots, a_N)^T$ must satisfy

$$\mathbf{a}^T \mathbf{m} = E\{S_k\}, \tag{4.16}$$

where $\mathbf{m} = (E\{x_{(1)}\}, E\{x_{(2)}\}, \ldots, E\{x_{(N)}\})^T$. We have arrived at a least squares problem with linear constraints. There are general methods of handling this type of problem [46, 99], but for our purposes the following simple result solves both of the cases above.

Let $\mathbf{R}$ be a positive definite $N \times N$ matrix, $\mathbf{r}$ and $\mathbf{b}$ $N \times 1$ vectors. Recall that a matrix $\mathbf{A}$ is positive definite if $\mathbf{x}^T \mathbf{A} \mathbf{x} > 0$ whenever $\mathbf{x} \neq \mathbf{0} = (0, 0, \ldots, 0)^T$. The solution of the minimization problem

$$\begin{aligned} \text{minimize } v(\mathbf{x}) &= \mathbf{x}^T \mathbf{R} \mathbf{x} - 2\mathbf{x}^T \mathbf{r} \\ \text{subject to } \mathbf{x}^T \mathbf{b} &= c \end{aligned} \tag{4.17}$$

is

$$\mathbf{x}_0 = \mathbf{R}^{-1} \mathbf{r} + \frac{c - \mathbf{b}^T \mathbf{R}^{-1} \mathbf{r}}{\mathbf{b}^T \mathbf{R}^{-1} \mathbf{b}} \mathbf{R}^{-1} \mathbf{b}. \tag{4.18}$$

That (4.18) is the solution of (4.17) can be seen as follows. Direct substitution shows that (4.18) satisfies the constraint (4.17). Now the function

$$
\begin{aligned}
v(\mathbf{y} + \mathbf{x}_0) &= (\mathbf{y} + \mathbf{x}_0)^T \mathbf{R}(\mathbf{y} + \mathbf{x}_0) - 2(\mathbf{y} + \mathbf{x}_0)^T \mathbf{r} \\
&= \mathbf{y}^T \mathbf{R} \mathbf{y} + 2\mathbf{y}^T (\mathbf{R}\mathbf{x}_0 - \mathbf{r}) + \text{constant} \\
&= \mathbf{y}^T \mathbf{R} \mathbf{y} - \left(\frac{c - \mathbf{b}^T \mathbf{R}^{-1} \mathbf{r}}{\mathbf{b}^T \mathbf{R}^{-1} \mathbf{b}} \right) \underbrace{\mathbf{y}^T \mathbf{b}}_{\text{constant}} + \text{constant}
\end{aligned}
$$

clearly has its minimum under the constraint (4.17) at zero because $\mathbf{R}$ is positive definite.

From (4.18) it immediately follows that the optimal coefficient vector $\mathbf{a}$ is

$$
\mathbf{a} = \mathbf{R}^{-1} \mathbf{r} + \frac{1 - \mathbf{1}^T \mathbf{R}^{-1} \mathbf{r}}{\mathbf{1}^T \mathbf{R}^{-1} \mathbf{1}} \mathbf{R}^{-1} \mathbf{1} \tag{4.19}
$$

for the condition (4.15) and

$$
\mathbf{a} = \mathbf{R}^{-1} \mathbf{r} + \frac{E\{S_k\} - \mathbf{m}^T \mathbf{R}^{-1} \mathbf{r}}{\mathbf{m}^T \mathbf{R}^{-1} \mathbf{m}} \mathbf{R}^{-1} \mathbf{m} \tag{4.20}
$$

for the condition (4.16).

If we in the case of location invariance further assume a constant signal and additive white zero mean noise, independent of the signal,

$$
X_n = S + N_n,
$$

the mean square error

$$
E\{(Y - S)^2\} = E\{(\sum a_n(S + N_{(n)}) - S)^2)\} = E\{(\sum a_n N_{(n)})^2)\} = \mathbf{a}^T \mathbf{R} \mathbf{a},
$$

where $\mathbf{R} = [E\{N_{(i)} N_{(j)}\}]$ is the correlation matrix of the noise. Now, we have $\mathbf{r} = \mathbf{0}$ in (4.17), and so (4.19) gives

$$
\mathbf{a} = \frac{\mathbf{R}^{-1} \mathbf{1}}{\mathbf{1}^T \mathbf{R}^{-1} \mathbf{1}}. \tag{4.21}
$$

Example 4.2.   Assume that we have a constant signal (of random amplitude) corrupted by i.i.d. additive noise uniformly distributed on $[-1/2, 1/2]$ and independent of the signal. Let us use (4.21) to compute the optimal coefficients of an L-filter of length N. We need the correlation matrix of the noise

$$
\mathbf{R} = \left[E\{N_{(i)} N_{(j)}\} \right],
$$

where $N_1, N_2, \ldots, N_N$ are i.i.d. and uniformly distributed on $[-1/2, 1/2]$. When $U_1, U_2, \ldots, U_N$ are independent and uniformly distributed on $[0, 1]$, using the Beta distribution and the joint distribution of two order statistics (4.6) direct integration yields (Exercise)

$$
E\{U_{(i)}\} = \frac{i}{N + 1}
$$

and

$$
E\{U_{(i)} U_{(j)}\} = \frac{ij + \min\{i, j\}}{(N + 1)(N + 2)},
$$

where $i, j = 1, 2, \ldots, N$. Now, we can write $N_i = U_i - \frac{1}{2}$ and so

$$E\{N_{(i)}N_{(j)}\} = E\{U_{(i)}U_{(j)}\} - \frac{1}{2}(E\{U_{(i)}\} + E\{U_{(j)}\}) + \frac{1}{4}$$

giving

$$\mathbf{R} = \left[\frac{4ij + 4\min\{i,j\} - 2(N+2)(i+j) + (N+1)(N+2)}{4(N+1)(N+2)}\right]. \tag{4.22}$$

Fortunately the inverse of $\mathbf{R}$ is fairly straightforward to obtain. Direct computation (Exercise) verifies that the inverse of $\mathbf{R}$ is

$$\begin{pmatrix} \frac{3N-2}{8N} & -\frac{1}{4} & 0 & \cdots & 0 & 0 & \frac{N+2}{8N} \\ -\frac{1}{4} & \frac{1}{2} & -\frac{1}{4} & \cdots & 0 & 0 & 0 \\ \vdots & \vdots & \vdots & \ddots & \vdots & \vdots & \vdots \\ 0 & 0 & 0 & \cdots & -\frac{1}{4} & \frac{1}{2} & -\frac{1}{4} \\ \frac{N+2}{8N} & 0 & 0 & \cdots & 0 & -\frac{1}{4} & \frac{3N-2}{8N} \end{pmatrix},$$

which finally gives

$$\mathbf{a} = (\frac{1}{2}, 0, \ldots, 0, \frac{1}{2})^T,$$

the midrange of $X_1, X_2, \ldots, X_N$.

Example 4.3. Suppose that we have a constant signal but that the level of the signal is a random variable the mean and variance of which are both equal to 1. Assume further that the signal is corrupted by i.i.d. noise with uniform $U(0,1)$ distribution and which is independent of the signal. Let us compute the optimal L-filter of length 5 that would be unbiased (even though we have nonsymmetric noise). Now,

$$\mathbf{r} = (E\{SX_{(1)}\}, E\{SX_{(2)}\}, \ldots, E\{SX_{(N)}\})^T = \frac{1}{6}(1, 2, 3, 4, 5)^T$$

and the unbiased condition gives

$$\mathbf{m}^T\mathbf{a} = \frac{1}{6}(7, 8, 9, 10, 11) \, \mathbf{a} = E\{S\} = 1.$$

The correlation matrix $\mathbf{R}$ is

$$\mathbf{R} = \left[E\{(U_{(i)} + S)(U_{(j)} + S)\}\right] = \left[E\{U_{(i)}U_{(j)}\} + E\{U_{(i)}S\} + E\{U_{(j)}S\} + E\{S^2\}\right].$$

Inverting $\mathbf{R}$ and substituting into (4.20) gives

$$\mathbf{a} = (-3.5939, 0, 0, 0, 2.8325)^T.$$

It is intuitively obvious that if we just computed the midrange and then subtracted $\frac{1}{2}$ from it the result would be much better. The fact that we try to compensate the nonsymmetric noise with a linear operation leads to the strange values.

Remark 4.2. The above methods apply to any filter that can be expressed as a linear combination of nonlinear filters provided that it is possible to compute the second order statistics of the outputs of nonlinear filters.

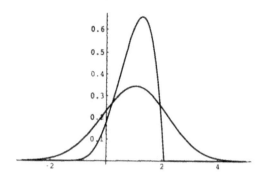

Figure 4.1. The output density of Example 4.3 (broader). The output density of Example 4.3 modified so that the uncorrupted signal is the nonrandom constant 1 (narrower).

4.2   Stack Filters

4.2.1   Output Distribution of a Stack Filter

The output of a stack filter at each window position is the result of a sum of binary operations operating on thresholded versions of the samples appearing in the window of the filter. The key to the analysis of stack filters comes from their definition by threshold decomposition [56, 112, 117, 121]. By threshold decomposition we can divide the analysis of stack filters into smaller and simpler parts. In other words, most of the analysis can be done by studying binary signals.

 Consider a vector $\mathbf{x} = (X_1, X_2, \ldots, X_N)$, where $X_i \in \{0, 1, \ldots, M-1\}$. The *threshold decomposition* of $\mathbf{x}$ means decomposing $\mathbf{x}$ into $M-1$ binary vectors $\mathbf{x}^{(1)}, \mathbf{x}^{(2)}, \ldots, \mathbf{x}^{(M-1)}$, according to the thresholding rule

$$x_n^{(m)} = T^{(m)}(X_n) = \begin{cases} 1, & \text{if } X_n \geq m, \\ 0, & \text{otherwise.} \end{cases} \qquad (4.23)$$

Thus, the binary vector $\mathbf{x}^{(m)}$ is obtained by thresholding the input vector at the level m, for $1 \leq m \leq M-1$. An element $x_n^{(k)}$ of the binary vector $\mathbf{x}^{(k)}$ takes on the value 1 whenever the element of the input vector x_n is greater than or equal to k.

Let us first suppose that the input signal is a sequence of binary random variables. Consider a particular time instant. We can assume that the window contains $X_1, X_2, \ldots, X_N$, where X_i are binary independent random variables and denote $P(X_i = 0) = p_i$, $i = 1, 2, \ldots, N$. Now, the output is $Y = f(X_1, X_2, \ldots, X_N)$, where $f(x_1, x_2, \ldots, x_N)$ is the positive Boolean function that defines the stack filter. The sample space is

$$\{0, 1\}^N = \{(x_1, x_2, \ldots, x_N) : x_i \in \{0, 1\}\}$$

and clearly

$$P\left((X_1, X_2, \ldots, X_N) = (x_1, x_2, \ldots, x_N)\right) = \prod_{i=1}^{N} p_i^{1-x_i}(1 - p_i)^{x_i}. \qquad (4.24)$$

The probability that the output has the value 0 is a union of elementary events which are necessarily disjoint. Thus,

$$
\begin{aligned}
P(Y = 0) &= \sum_{f(\mathbf{x})=0} \prod_{i=1}^{N} p_i^{1-x_i}(1 - p_i)^{x_i} \\
&= \sum_{\mathbf{x} \in f^{-1}(0)} \prod_{i=1}^{N} p_i^{1-x_i}(1 - p_i)^{x_i},
\end{aligned}
\tag{4.25}
$$

where $f^{-1}(0)$ is the *pre-image* of 0, i.e., $f^{-1}(0) = \{\mathbf{x} : f(\mathbf{x}) = 0\}$, and the binary values in the exponents are to be understood as real 0's and 1's.

Remark 4.3. Note that the positivity of the defining Boolean function appears nowhere in the derivation of (4.25), which means that in the case of a binary signal, (4.25) holds also for threshold Boolean filters. The positivity of the Boolean function is the key element in extending (4.25) to the nonbinary case.

Remark 4.4. If the binary signal above is "white noise" in the sense that the random variables X_n have identical distributions in addition to being independent so that $P(X_i = 0) = p$, $i = 1, 2, \ldots, N$, then (4.25) gets an even simpler form:

$$
\begin{aligned}
P(Y = 0) &= \sum_{f(\mathbf{x})=0} \prod_{i=1}^{N} p^{N-w(\mathbf{x})}(1 - p)^{w(\mathbf{x})} \\
&= \sum_{\mathbf{x} \in f^{-1}(0)} \prod_{i=1}^{N} p^{N-w(\mathbf{x})}(1 - p)^{w(\mathbf{x})},
\end{aligned}
$$

where $w(\mathbf{x})$ is the *Hamming weight* of $\mathbf{x}$, i.e., the number of components equal to 1 in $\mathbf{x}$.

Continuous Stack Filters

We saw earlier that irrespective of the actual values of the threshold levels of the signal the output of the stack filter that is defined by the positive Boolean function $f(x_1, x_2, \ldots, x_N)$ can, for the input $\mathbf{x} = (X_1, X_2, \ldots, X_N)$, be given by

$$
S_f(\mathbf{x}) = \text{MAX}\{\text{MIN}\{X_j : j \in P_1\}, \text{MIN}\{X_j : j \in P_2\}, \ldots, \text{MIN}\{X_j : j \in P_K\}\}. \tag{4.26}
$$

Thus, the *continuous stack filter* corresponding to a positive Boolean function can be expressed by replacing AND and OR with MIN and MAX, respectively. For example, the three-point median filter over real variables X_1, X_2, and X_3 is the continuous stack filter defined by the positive Boolean function $f(x_1, x_2, x_3) = x_1 x_2 + x_1 x_3 + x_2 x_3$, i.e.,

$$
\text{MED}\{X_1, X_2, X_3\} = \text{MAX}\{\text{MIN}\{X_1, X_2\}, \text{MIN}\{X_1, X_3\}, \text{MIN}\{X_2, X_3\}\}.
$$

The output distribution of a stack filter when the inputs are independent and have continuous distributions is readily derived using thresholding and the same reasoning as in the binary case. Let the input values $X_1, X_2, \ldots, X_N$ of a stack filter $S_f(\cdot)$ defined by a positive Boolean function $f(x_1, x_2, \ldots, x_N)$ be independent random variables having the distribution functions $\Phi_1(t), \Phi_2(t), \ldots, \Phi_N(t)$, respectively. Then the output distribution function $\Psi(t)$ of the continuous stack filter is

$$
\Psi(t) = \sum_{\mathbf{x} \in f^{-1}(0)} \prod_{i=1}^{N} (1 - \Phi_i(t))^{x_i} \Phi_i(t)^{1-x_i}. \tag{4.27}
$$

This can be seen as follows. Let Y be the output (random variable) of the filter. Then

$$\Psi(t) = P\{Y \leq t\}.$$

The input space can be divided into 2^N mutually disjoint events

$$E_{\mathbf{u}_0} = (-\infty, t] \times (-\infty, t] \times \cdots \times (-\infty, t],$$
$$E_{\mathbf{u}_1} = (t, \infty) \times (-\infty, t] \times \cdots \times (-\infty, t],$$
$$\vdots$$
$$E_{\mathbf{u}_{2^N}} = (t, \infty) \times (t, \infty) \times \cdots \times (t, \infty),$$

where the coding of events with binary vectors $\mathbf{u} = (u_1, u_2, \ldots, u_N)$ is such that

$$\begin{cases} u_b = 1, & \text{corresponds to} \quad (t, \infty), \\ u_b = 0, & \text{corresponds to} \quad (-\infty, t]. \end{cases}$$

Now, it is straightforward to verify that $Y \leq t$ if and only if the input vector $\mathbf{x}$ belongs to an event $E_{\mathbf{u}}$ such that $f(\mathbf{u}) = 0$. The probability of such an event $E_{\mathbf{u}}$ is

$$\prod_{i=1}^{N} (1 - \Phi_i(t))^{u_i} \Phi_i(t)^{1-u_i}.$$

Thus

$$\Psi(t) = \sum_{\mathbf{u} \in f^{-1}(0)} \prod_{i=1}^{N} (1 - \Phi_i(t))^{u_i} \Phi_i(t)^{1-u_i}.$$

In the case of independent and identically distributed input values (4.27) implies the following result.

Let the input values $X_1, X_2, \ldots, X_N$, in the window of the stack filter $S_f(\cdot)$ defined by a positive Boolean function $f(\cdot)$ be independent, identically distributed random variables having a common distribution function $\Phi(t)$. Then the distribution function $\Psi(t)$ of the output of the stack filter $S_f(\cdot)$ is

$$\Psi(t) = \sum_{i=0}^{N} A_i (1 - \Phi(t))^i \Phi(t)^{N-i}, \tag{4.28}$$

where the numbers A_i are defined by

$$A_i = |\{\mathbf{x} : f(\mathbf{x}) = 0, w_H(\mathbf{x}) = i\}|, \tag{4.29}$$

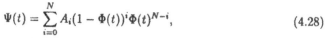

$|\Omega|$ denotes the cardinality of the set $|\Omega|$, and $w_H(\mathbf{x})$ is the Hamming weight of $\mathbf{x}$.

In order to guarantee that the stack filter is not defined by the trivial positive Boolean functions $f(\mathbf{x}) = 0$ for all $\mathbf{x}$, or $f(\mathbf{x}) = 1$ for all $\mathbf{x}$, we require

$$A_0 = 1 \quad \text{and} \quad A_N = 0. \tag{4.30}$$

As $A_N = 0$, we can leave out $i = N$ from the sum (4.28).

Remark 4.5. Because the number A_i is the cardinality of a set of binary vectors of length N they obviously satisfy

$$0 \leq A_i \leq \binom{N}{i}, \quad i = 1, 2, \ldots, N.$$

Example 4.4. All ranked order filters are stack filters. For instance, the positive Boolean function corresponding to the rth order statistic can be given as

$$f_r(x_1, x_2, \ldots, x_N) = \begin{cases} 1, & \text{if } x_1 + x_2 + \cdots + x_N \geq N - r + 1, \\ 0, & \text{otherwise,} \end{cases} \quad (4.31)$$

where the sum on the right-hand side of (4.31) is real.

Remark 4.6. Boolean functions that can be expressed in the form

$$f(x_1, x_2, \ldots, x_N) = \begin{cases} 1, & \text{if } w_1 x_1 + w_2 x_2 + \cdots + w_N x_N \geq t, \\ 0, & \text{otherwise} \end{cases}$$

are called *threshold functions* or *linearly separable Boolean functions*. The theory of threshold functions, *threshold logic*, gives efficient methods of realizing switching functions and was extensively studied in the sixties (c.f. Reference [77]).

Example 4.5. Let $X_1, X_2, \ldots, X_N$ be i.i.d. random variables with a common distribution function $\Phi(t)$. Let Y be the output of the continuous stack filter defined by (4.31). Now

$$f_r(x_1, x_2, \ldots, x_N) = \begin{cases} 1, & \text{if } x_1 + x_2 + \cdots + x_N \geq N - r + 1, \\ 0, & \text{otherwise} \end{cases}$$

implying that

$$A_i = \begin{cases} \binom{N}{i}, & i = 1, 2, \ldots, N - r, \\ 0, & i = N - r + 1, N - r + 2, \ldots, N, \end{cases}$$

and by formula (4.28) the output distribution function of Y is

$$\begin{aligned} \Psi(t) &= \sum_{i=1}^{N} A_i (1 - \Phi(t))^i \Phi(t)^{N-i} \\ &= \sum_{i=0}^{N-r} \binom{N}{i} (1 - \Phi(t))^i \Phi(t)^{N-i} \\ &= \sum_{i=0}^{N-r} \binom{N}{N-i} (1 - \Phi(t))^i \Phi(t)^{N-i} \\ &\underset{j=N-i}{=} \sum_{j=r}^{N} \binom{N}{j} (1 - \Phi(t))^{N-j} \Phi(t)^j \end{aligned} \quad (4.32)$$

which corresponds to the density (4.5).

Example 4.6. Consider the stack filter defined by the Boolean function $f(x_1, x_2, x_3, x_4) = x_1 x_2 + x_1 x_3 + x_2 x_3 + x_2 x_4 + x_3 x_4$. Direct computation shows that

$$A_0 = 1, \quad A_1 = 4, \quad A_2 = 1, \text{ and } A_i = 0, \text{ for } i = 3, 4.$$

Assuming that the inputs are i.i.d. with the common distribution function $\Phi(t)$, the output distribution is

$$\Psi(t) = \Phi(t)^4 + 4(1 - \Phi(t))\Phi(t)^3 + (1 - \Phi(t))^2 \Phi(t)^2 = \Phi(t)^2 + 2\Phi(t)^3 - 2\Phi(t)^4. \quad (4.33)$$

The input and output densities are plotted in Figure 4.2. The input density is the Cauchy density. Note that the output density is more peaked and narrower which reflects the fact that this stack filter attenuates noise. Notice also that this Boolean function does not contain all products of two terms which means that it favors large values. This is reflected in the fact that the output density has shifted towards the right.

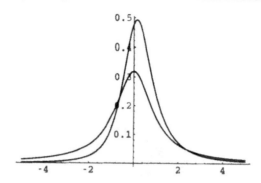

Figure 4.2. Input and output densities of Example 4.6.

Example 4.7. (The statistical symmetry of self-dual stack filters.)   The *dual of a Boolean function* $f(x_1, x_2, \ldots, x_N)$ is defined as

$$f^D(x_1, x_2, \ldots, x_N) = \overline{f(\overline{x_1}, \overline{x_2}, \ldots, \overline{x_N})},$$

i.e., it is the Boolean function that is obtained from f by first complementing all the variables and then the result. Equivalently the dual is obtained from the sum of products form by changing each sum to product and vice versa. A Boolean function f is called *self-dual* if it equals its dual, i.e.,

$$f(x_1, x_2, \ldots, x_N) = f^D(x_1, x_2, \ldots, x_N) \quad \text{for all} \quad (x_1, x_2, \ldots, x_N) \in \{0, 1\}^N.$$

The Boolean function $f(x_1, x_2, x_3) = x_1 x_2 + x_1 x_3 + x_2 x_3$ is self-dual. This can be seen either directly from the definition by checking each value in the table below

x_1	x_2	x_3	f	$\overline{x_1}$	$\overline{x_2}$	$\overline{x_3}$	f	$\overline{f}$
0	0	0	0	1	1	1	1	0
0	0	1	0	1	1	0	1	0
0	1	0	0	1	0	1	1	0
0	1	1	1	1	0	0	0	1
1	0	0	0	0	1	1	1	0
1	0	1	1	0	1	0	0	1
1	1	0	1	0	0	1	0	1
1	1	1	1	0	0	0	0	1,

or by the manipulation

$$f^D(x_1, x_2, \ldots, x_N) = (x_1 + x_2)(x_1 + x_3)(x_2 + x_3)$$
$$= x_1 x_1 x_2 + x_1 x_1 x_3 + x_1 x_3 x_2 + x_1 x_3 x_3 + x_2 x_1 x_2 + x_2 x_1 x_3 + x_2 x_3 x_2 + x_2 x_3 x_3$$
$$= x_1 x_2 + x_1 x_3 + x_2 x_3.$$

A stack filter is called *self-dual* if the positive Boolean function that defines it is self-dual. Consider now a stack filter $S_f(\cdot)$ defined by a self-dual positive Boolean function $f(x_1, x_2, \ldots, x_N)$ and assume that the distribution functions $\Phi_i(t)$ of the independent input random variables X_i are symmetrical with respect to the origin, i.e., $\Phi_i(t) = 1 - \Phi_i(-t)$. Then also the output random variable $S_f(X_1, X_2, \ldots, X_N)$ is symmetrical with respect to the origin. Using the fact that

$$\overline{\mathbf{x}} \in f^{-1}(0) \quad \text{if and only if} \quad \mathbf{x} \in f^{-1}(1),$$

this can be seen as follows. By formula (4.27) we have

$$\Psi(t) + \Psi(-t) = \sum_{\mathbf{x} \in f^{-1}(0)} \prod_{i=1}^{N}(1 - \Phi_i(t))^{x_i}\Phi_i(t)^{1-x_i} + \sum_{\mathbf{x} \in f^{-1}(0)} \prod_{i=1}^{N}(1 - \Phi_i(-t))^{x_i}\Phi_i(-t)^{1-x_i}$$

$$\underset{\Phi_i(t)=1-\Phi_i(-t)}{=} \sum_{\mathbf{x} \in f^{-1}(0)} \prod_{i=1}^{N}(1 - \Phi_i(t))^{x_i}\Phi_i(t)^{1-x_i} + \sum_{\mathbf{x} \in f^{-1}(0)} \prod_{i=1}^{N}\Phi_i(t)^{x_i}(1 - \Phi_i(-t))^{1-x_i}$$

$$= \sum_{\mathbf{x} \in f^{-1}(0)} \prod_{i=1}^{N}(1 - \Phi_i(t))^{x_i}\Phi_i(t)^{1-x_i} + \sum_{\overline{\mathbf{x}} \in f^{-1}(0)} \prod_{i=1}^{N}\Phi_i(t)^{\overline{x_i}}(1 - \Phi_i(t))^{1-\overline{x_i}}$$

$$= \sum_{\mathbf{x} \in f^{-1}(0)} \prod_{i=1}^{N}(1 - \Phi_i(t))^{x_i}\Phi_i(t)^{1-x_i} + \sum_{\mathbf{x} \in f^{-1}(1)} \prod_{i=1}^{N}(1 - \Phi_i(t))^{x_i}\Phi_i(t)^{1-x_i}$$

$$= \sum_{\mathbf{x} \in \{0,1\}^N} \prod_{i=1}^{N}(1 - \Phi_i(t))^{x_i}\Phi_i(t)^{1-x_i} = \prod_{i=1}^{N}((1 - \Phi_i(t)) + \Phi_i(t)) = 1.$$

The last equality above is just an equality of the type $(a_1 + b_1)(a_2 + b_2) = a_1^1 b_1^0 a_2^1 b_2^0 + a_1^1 b_1^0 a_2^0 b_2^1 + a_1^0 b_1^1 a_2^1 b_2^0 + a_1^0 b_1^0 a_2^0 b_2^1$. It immediately follows from the symmetry of the output distribution that the expectation of the output of a self-dual stack filter is zero if the input is symmetrical zero mean noise. We will use this result later when we consider optimization of stack filters.

4.2.2   Joint Distribution of Two Stack Filters

The methods that were used in Section 4.1.2 to optimize L-filters apply equally well to any filter that is formed as a linear combination of stack filters. To form the correlation matrix we need to compute the correlations between the outputs of the stack filters. This can be done if we know the joint distribution functions of the stack filters. In the following we derive a formula for the joint distribution function of two stack filters [1, 122].

Let $f(x_1, x_2, \ldots, x_N)$ and $g(x_1, x_2, \ldots, x_N)$ be two positive Boolean functions and $S_f(\cdot)$ and $S_g(\cdot)$ be the corresponding continuous stack filters. Let $X_1, X_2, \ldots, X_N$ be independent random variables with distribution functions $\Phi_1(t), \Phi_2(t), \ldots, \Phi_N(t)$, respectively. Denote the random variables formed by the outputs of the stack filters $S_f(\cdot)$ and $S_g(\cdot)$ by Y and Z, respectively. The joint distribution function of Y and Z is

$$H(s,t) = P(Y \le s, Z \le t).$$

We will use the same idea as when deriving the distribution function of a single stack filter. Assume for the time being that $s \le t$ and divide $\mathbf{R}$ into three disjoint intervals:

$$V_0 = (-\infty, s],$$
$$V_1 = (s, t],$$
$$V_2 = (t, \infty).$$

Then, because $\mathbf{R}^N = (V_0 \cup V_1 \cup V_2)^N$, we can write $\mathbf{R}^N$ as the disjoint union

$$\mathbf{R}^N = \bigcup V_{a_1} \times V_{a_2} \times \cdots \times V_{a_N} = \bigcup_{\mathbf{a} \in \mathcal{I}} V(\mathbf{a}),$$

where $\mathbf{a} = (a_1, a_2, \ldots, a_N)$ runs through the 3^N elements of $\mathcal{I} = \{(a_1, a_2, \ldots, a_N) : a_i \in \{0,1,2\}\} = \{0,1,2\}^N$. The partition of $\mathbf{R}^2$ into the sets $V(\mathbf{a})$ is illustrated in Figure 4.3.

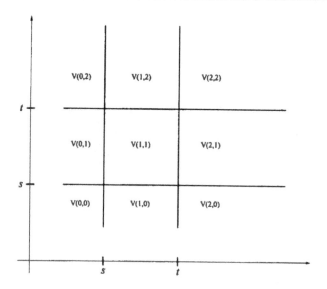

Figure 4.3. The division of $\mathbf{R}^2$ into 9 two-dimensional intervals.

The key observation is that the value of Y can cross the point s only if some of the signal values X_i cross the point $X_i = s$. Similarly the value of Z can cross the point t only if some of the signal values X_i cross the point $X_i = t$. Thus, the relation $(Y \leq s, Z \leq t)$ is either true or not true uniformly in each set $V(a_1, a_2, \ldots, a_N)$. To get the probability $P(Y \leq s, Z \leq t)$ we only need to compute the probability of the event corresponding to the union of the favorable sets $V(\mathbf{a})$. Because the sets are disjoint this is the sum of the probabilities of the favorable sets.

Consider a set $V(a_1, a_2, \ldots, a_N)$. The probability that the component X_i is in V_{a_i} is

$$P(X_i \in V_{a_i}) = \begin{cases} \Phi_i(s), & \text{if } a_i = 0, \\ \Phi_i(t) - \Phi_i(s), & \text{if } a_i = 1, \\ 1 - \Phi_i(t), & \text{if } a_i = 2, \end{cases}$$

and because of independence

$$P(X \in V(\mathbf{a})) = \prod_{i=1}^{N} P(x_i \in V_{a_i}).$$

The final problem is to find an indexing scheme that would make it easy to keep track of the subsets $V(\mathbf{a})$ and the values of the stack filters $S_f(\cdot)$ and $S_g(\cdot)$ simultaneously. To achieve this we rewrite $\mathcal{I} = \{(a_1, a_2, \ldots, a_N) : a_i \in \{0, 1, 2\}\}$ as

$$\mathcal{I} = \{(b_1, b_2, \ldots, b_N) + (c_1, c_2, \ldots, c_N) : b_i, c_i \in \{0, 1\}, c_i \leq b_i\},$$

where the sum is real. This is clarified by the following simple example.

Example 4.8. Let $N = 2$. Then we have the decomposition

$$
\begin{aligned}
(0,0) &= (0,0) + (0,0) \\
(0,1) &= (0,1) + (0,0) \\
(0,2) &= (0,1) + (0,1) \\
(1,0) &= (1,0) + (0,0) \\
(1,1) &= (1,1) + (0,0) \\
(1,2) &= (1,1) + (0,1) \\
(2,0) &= (1,0) + (1,0) \\
(2,1) &= (1,1) + (1,0) \\
(2,2) &= (1,1) + (1,1).
\end{aligned}
$$

Now, consider a fixed realization $\mathbf{x} = (x_1, x_2, \ldots, x_N)$ of $(X_1, X_2, \ldots, X_N)$. Let $\mathbf{b}$ be $\mathbf{x}$ thresholded at s and $\mathbf{c}$ be $\mathbf{x}$ thresholded at t, i.e.,

$$
\mathbf{b} = T^{(s)}(\mathbf{x}),
$$
$$
\mathbf{c} = T^{(t)}(\mathbf{x}).
$$

The outputs of $S_f(\cdot)$ and $S_g(\cdot)$ satisfy

$$
S_f(\mathbf{x}) \le s \text{ if and only if } f(\mathbf{b}) = 0,
$$
$$
S_g(\mathbf{x}) \le t \text{ if and only if } g(\mathbf{c}) = 0,
$$

from which we can now immediately write the probability of the event $(Y \le s, Z \le t)$ where $s \le t$ as

$$
P(Y \le s, Z \le t) = \sum_{\substack{\mathbf{c} \le \mathbf{b} \\ f(\mathbf{b}) = g(\mathbf{c}) = 0}} \prod_{i=1}^{N} \Phi_i(s)^{(1-b_i)(1-c_i)} \big(\Phi_i(t) - \Phi_i(s)\big)^{b_i(1-c_i)} \big(1 - \Phi_i(t)\big)^{b_i c_i},
$$

(4.34)

since

$$
\begin{aligned}
(1 - b_i)(1 - c_i) &= \begin{cases} 1, & \text{if } b_i = c_i = 0, \text{ i.e., } a_i = 0, \\ 0, & \text{otherwise,} \end{cases} \\
b_i(1 - c_i) &= \begin{cases} 1, & \text{if } b_i = 1, \ c_i = 0, \text{ i.e., } a_i = 1, \\ 0, & \text{otherwise,} \end{cases} \\
b_i c_i &= \begin{cases} 1, & \text{if } b_i = c_i = 1, \text{ i.e., } a_i = 2, \\ 0, & \text{otherwise.} \end{cases}
\end{aligned}
$$

For $s > t$ we just need to reverse the roles of $\mathbf{b}$ and $\mathbf{c}$ and we obtain

$$
P(Y \le s, Z \le t) = \sum_{\substack{\mathbf{b} \le \mathbf{c} \\ f(\mathbf{b}) = g(\mathbf{c}) = 0}} \prod_{i=1}^{N} \Phi_i(t)^{(1-b_i)(1-c_i)} \big(\Phi_i(s) - \Phi_i(t)\big)^{b_i(1-c_i)} \big(1 - \Phi_i(s)\big)^{b_i c_i}.
$$

(4.35)

The formulas (4.34) and (4.35) together give the joint distribution function of two stack filters.

Remark 4.7. If we set $s = t$ in (4.34) and (4.35) they collapse to the distribution function of the maximum of $S_f(\cdot)$ and $S_g(\cdot)$.

Example 4.9. Let $f(x_1, x_2) = x_1$ and $g(x_1, x_2) = x_1 x_2$. Thus $S_f(X_1, X_2) = X_1$ and $S_g(X_1, X_2) = \text{MIN}\{X_1, X_2\}$. Assume that both X_1 and X_2 have the distribution function $\Phi(t)$ and let first $s \leq t$. Then the pairs **b** and **c** satisfying

$$\mathbf{b} \geq \mathbf{c}, \quad f(\mathbf{b}) = 0, \quad g(\mathbf{c}) = 0$$

are

b	**c**
$(0,0)$	$(0,0)$
$(0,1)$	$(0,0)$
$(0,1)$	$(0,1)$

giving

$$\Psi(s,t) = P(Y \leq s, Z \leq t)$$
$$= \Phi(t)\Phi(s) + \Phi(s)(\Phi(t) - \Phi(s)) + \Phi(s)(1 - \Phi(t))$$
$$= \Phi(s).$$

Let then $s > t$. The pairs **b** and **c** satisfying

$$\mathbf{b} \leq \mathbf{c}, \quad f(\mathbf{b}) = 0, \quad g(\mathbf{c}) = 0$$

are

b	**c**
$(0,0)$	$(0,0)$
$(0,0)$	$(0,1)$
$(0,0)$	$(1,0)$
$(0,1)$	$(0,1)$

giving

$$\Psi(s,t) = \Phi(t)\Phi(s) + \Phi(t)(\Phi(s) - \Phi(t)) + \Phi(t)(\Phi(s) - \Phi(t)) + \Phi(t)(1 - \Phi(s))$$
$$= \Phi(t)(1 + \Phi(s) - \Phi(t)).$$

Thus

$$\Psi(s,t) = \begin{cases} \Phi(s), & \text{if } s \leq t, \\ \Phi(t)(1 + \Phi(s) - \Phi(t)), & \text{if } s > t. \end{cases}$$

Remark 4.8. The methods above generalize to more than two stack filters. If we have L stack filters the space $\mathbf{R}^N$ gets, in principle, divided into $(L+1)^N$ N-dimensional intervals, and efficient bookkeeping of the favorable events becomes more difficult.

Remark 4.9. In the case of i.i.d. input, that is, $\Phi_i(t) = \Phi(t)$ for all i, and t we can rewrite the formula (4.34) as

$$\Psi(s,t) = \sum_{i=0}^{N} \sum_{j=0}^{N} A_{ij} \Phi(s)^i (\Phi(t) - \Phi(s))^{N-i-j} (1 - \Phi(t))^j, \qquad (4.36)$$

where

$$A_{ij} = |\{(\mathbf{b}, \mathbf{c}) \in \{0,1\}^N \times \{0,1\}^N : \mathbf{b} \geq \mathbf{c}, \ f(\mathbf{b}) = g(\mathbf{c}) = 0,$$
$$\sum_{k=1}^{N}(1 - b_k)(1 - c_k) = i, \ \sum_{k=1}^{N} b_k c_k = j\}|$$

for $i, j = 0, 1, \ldots, N$.

4.2.3 Output Moments of Stack Filters

In the theory of linear systems the power spectrum of the input signal together with the transfer function of the system determines the power spectrum of the output signal. It is not possible to get equally simple and strong connections between the input process and the output process for stack filters. Explicit relations between the input and output statistics can in practice only be derived for the case of a constant signal plus noise. Even then we need to assume the noise to be white. However, these relations enable us to numerically optimize noise attenuation of stack filters under constraints guaranteeing that the filter satisfies prescribed specifications.

We first derive a method of calculating the output moments of stack filters by using the coefficients A_i. We saw earlier that if the input values $X_1, X_2, \ldots, X_N$, in the window of the stack filter $S_f(\cdot)$ defined by a positive Boolean function $f(\cdot)$ are independent, identically distributed random variables having a common distribution function $\Phi(t)$, then the distribution function of the output $\Psi(t)$ of the stack filter $S_f(\cdot)$ is

$$\Psi(t) = \sum_{i=0}^{N} A_i (1 - \Phi(t))^i \Phi(t)^{N-i}, \qquad (4.37)$$

where the numbers A_i were defined by

$$A_i = |\{\mathbf{x} : f(\mathbf{x}) = 0, w_H(\mathbf{x}) = i\}|.$$

It immediately follows that the kth-order moment about the origin of the output Y of a stack filter can be expressed as

$$\mu'_k = E\{Y^k\} = \sum_{i=0}^{N-1} A_i M(\Phi, k, N, i), \qquad (4.38)$$

where

$$M(\Phi, k, N, i) = \int_{-\infty}^{\infty} x^k \frac{d}{dx} \left((1 - \Phi(x))^i \Phi(x)^{N-i} \right) dx, \quad i = 0, 1, \ldots, N-1. \qquad (4.39)$$

By using the output moments about the origin we easily obtain *output central moments*, denoted by $\mu_k = E\left\{(Y - E\{Y\})^k\right\}$. For example, the second order central output moment equals

$$\mu_2 = \sum_{i=0}^{N-1} A_i M(\Phi, 2, N, i) - \left(\sum_{i=0}^{N-1} A_i M(\Phi, 1, N, i) \right)^2. \qquad (4.40)$$

The output variance is a measure of the noise attenuation capability of a filter. It quantifies the spread of the output samples with respect to their mean value. Equation (4.40) gives an expression for the output variance in terms of the quantities $M(\Phi, 1, N, i)$, $M(\Phi, 2, N, i)$, and the coefficients A_i. The important factor is that $M(\Phi, k, N, i)$ depend only on the input distribution Φ and the window size N, but not on the stack filter in question. The coefficients A_i, on the other hand, depend only on the stack filter and not on the input distribution.

In the following we discuss the properties of the numbers $M(\Phi, k, N, i)$. These properties will be used later when we present a method of optimizing stack filters. To avoid technical difficulties that might obscure the main ideas we assume that the input distribution $\Phi(t)$ is very smooth. Specifically, we assume that the density function $\phi(t)$ is positive for all t, symmetric with respect to the origin, and that

$$\lim_{t \to -\infty} t^2 \Phi(t) = \lim_{t \to \infty} t^2 (1 - \Phi(t)) = 0.$$

It follows from these assumptions that the expectation of the input is zero, i.e.,

$$\mu_x = 0,$$

and

$$\Phi(t) = 1 - \Phi(-t).$$

Even though the numbers $M(\Phi, k, N, i)$ are defined by integrals they are quite easy to compute numerically because they satisfy the following recurrence formula

$$M(\Phi, k, N, i) = M(\Phi, k, N-1, i-1) - M(\Phi, k, N, i-1), \quad 1 \le i \le N \qquad (4.41)$$

with initial values

$$M(\Phi, k, N, 0) = \int_{-\infty}^{\infty} x^k \frac{d}{dx}\left(\Phi(x)^N\right) dx, \quad i = 0, 1, \ldots, N-1, \quad 0 \le N.$$

This recurrence follows directly from the definition of the numbers $M(\Phi, k, N, i)$ by expanding in the integral (4.39) (Exercise)

$$(1 - \Phi(x))^i \Phi(x)^{N-i} = (1 - \Phi(x))^{i-1}\Phi(x)^{N-i} - (1 - \Phi(x))^{i-1}\Phi(x)^{N-i+1}.$$

To be able to minimize the output variance of a stack filter we need more detailed knowledge about the quantities $M(\Phi, 2, N, i)$ (and in general also $M(\Phi, 1, N, i)$) that appear in the formula for the output variance.

The symmetry of the input distribution implies that

$$(1 - \Phi(-x))^i \Phi(-x)^{N-i} = (1 - \Phi(x))^{N-i}\Phi(x)^i$$

and, thus

$$
\begin{aligned}
M(\Phi, 2, N, N-i) &= \int_{-\infty}^{\infty} x^2 \frac{d}{dx}\left((1 - \Phi(x))^{N-i}\Phi(x)^i\right) dx \\
&\underset{\Phi(-x)=1-\Phi(x)}{=} \int_{-\infty}^{\infty} x^2 \frac{d}{dx}\left((1 - \Phi(-x))^i\Phi(-x)^{N-i}\right) dx \\
&= -\int_{-\infty}^{\infty} x^2 \frac{d}{dx}\left((1 - \Phi(x))^i\Phi(x)^{N-i}\right) dx \\
&= -M(\Phi, 2, N, i).
\end{aligned}
\qquad (4.42)
$$

From (4.42) it follows that it is enough to consider $M(\Phi, 2, N, i)$ only in the range $0 \le i \le N/2$. First notice that if $i = 0$ the integrand is positive for all t and so $M(\Phi, 2, N, i) > 0$. Let $0 < i \le N/2$. Using integration by parts we get

$$
\begin{aligned}
M(\Phi, 2, N, i) &= \int_{-\infty}^{\infty} x^2 \frac{d}{dx}\left((1 - \Phi(x))^i\Phi(x)^{N-i}\right) dx \\
&= x^2(1 - \Phi(x))^i\Phi(x)^{N-i}\Big|_{-\infty}^{\infty} - 2\int_{-\infty}^{\infty} x(1 - \Phi(x))^i\Phi(x)^{N-i}dx \\
&= -2\left(\int_{-\infty}^{0} x(1 - \Phi(x))^i\Phi(x)^{N-i}dx + \int_{0}^{\infty} x(1 - \Phi(x))^i\Phi(x)^{N-i}dx\right) \\
&\underset{\Phi(-x)=1-\Phi(x)}{=} 2\left(\int_{0}^{\infty} x(1 - \Phi(x))^{N-i}\Phi(x)^i dx - \int_{0}^{\infty} x(1 - \Phi(x))^i\Phi(x)^{N-i}dx\right) \\
&= 2\int_{0}^{\infty} x\left((1 - \Phi(x))^{N-i}\Phi(x)^i - (1 - \Phi(x))^i\Phi(x)^{N-i}\right) dx.
\end{aligned}
$$

On the interval $(0, \infty)$ we have $1/2 < \Phi(x) < 1$ which implies that on the whole interval $(0, \infty)$

$$((1 - \Phi(x))^{N-i} \Phi(x)^i - (1 - \Phi(x))^i \Phi(x)^{N-i}) \quad \begin{cases} = 0, & \text{if } i = N/2, \\ < 0, & \text{if } 0 < i < N/2. \end{cases}$$

Combining with the case $i = 0$, this implies that $M(\Phi, 2, N, i)$ satisfy

$$M(\Phi, 2, N, i) \begin{cases} = 0, & i = N/2, \ (N \text{ even}) \\ > 0, & i = 0 \text{ or } N/2 < i < N, \\ < 0, & \text{otherwise}. \end{cases}$$

Assume that we wish to find the self-dual stack filter of length N (N odd) that best attenuates i.i.d. symmetric noise. Because the expectation of the output is zero the output variance is, according to (4.38),

$$\mu_2 = \sum_{i=0}^{N-1} A_i M(\Phi, 2, N, i).$$

From the above it follows that always $A_0 = 1$, $A_N = 0$, $M(\Phi, 2, N, i) < 0$ for $1 \leq i \leq (N-1)/2$ and $M(\Phi, 2, N, i) > 0$ for $(N+1)/2 \leq i \leq N-1$. This means that we must make A_i as large as possible for $1 \leq i \leq (N-1)/2$ and as small as possible for $(N+1)/2 \leq i \leq N-1$. This obviously happens if we choose $A_i = \binom{N}{i}$ for $1 \leq i \leq (N-1)/2$ and $A_i = 0$ for $(N+1)/2 \leq i \leq N-1$. This choice gives the median filter!

In a more meaningful situation we have additional constraints on the coefficients A_i which arise, e.g., from requirements that the filter must have a certain degree of robustness and it must also be able to preserve details of a prescribed type. The constraints that give detail preservation can be written down directly as predetermined values of the defining Boolean function. The robustness requirements are best described in terms of so called rank selection probabilities that are described in the next section.

4.2.4 Rank Selection Probabilities of Stack Filters

Many of the filters that were introduced in Chapter 3 had the property that the output of the filter is always one of the inputs in the filter window. Median filter, all stack filters, and permutation filters are examples of this type of filter for which one can define the probability that the output is of a fixed rank among the samples in the filter window, and also the probability that the output is the sample in a fixed position of the filter window. It turns out that the output distribution of the filter can be expressed in terms of these rank selection probabilities, and this will lead to interesting relations between the rank selection probabilities and the coefficients A_i that appeared in the formulas for the output distribution of a stack filter. Rank selection probabilities also give a natural way of expressing robustness constraints for filters. We can, for instance, require that the extreme ranks can never appear as outputs. This is conceptually quite similar to the idea of trimmed means even though the methods and the filters themselves come from different worlds.

Let $\mathbf{x} = (X_1, X_2, \ldots, X_N)$ denote a real-valued random vector in the input window of a filter $S(\cdot)$ that we assume to have the property that the output is always one of the inputs. We suppose that the random variables X_i are independent and identically distributed. Again, the corresponding order statistics are denoted by $X_{(i)}$, $i = 1, 2, \ldots, N$, i.e.,

$$X_{(1)} \leq X_{(2)} \leq \cdots \leq X_{(N)}.$$

Consider the probability that the ith smallest sample of the window is the output, and the probability that the jth sample of the window is the output. The random variables $X_1, X_2, \ldots, X_N$ may appear in $N!$ different orders, i.e., $((1), (2), \ldots, (N))$ can be any of the $N!$ permutations of $(1, 2, \ldots, N)$, where each permutation is called an ordering. Let z_k, $k = 1, 2, \ldots, N!$ represent such orderings. For simplicity we may assume that the probability that $X_i = X_j$ for some i, j, $i \neq j$ is zero so that these cases need not be considered. In general, dropping this assumption only causes slight notational difficulties. The ith *rank selection probability* is denoted by $P(Y = X_{(i)})$, $1 \leq i \leq N$, and is the probability that the output $Y = X_{(i)}$. The *rank selection probability vector* is the row vector $\mathbf{r} = (r_1, r_2, \ldots, r_N)$, where $r_i = P(Y = X_{(i)})$, $1 \leq i \leq N$. The jth *sample selection probability* is denoted by $P(Y = X_j)$, $1 \leq j \leq N$, and is the probability that the output $Y = X_j$. The *sample selection probability vector* is the row vector $\mathbf{s} = (s_1, s_2, \ldots, s_N)$, where $s_j = P(Y = X_j)$, $1 \leq j \leq N$.

We denote the number of distinct orderings for which $Y = X_{(i)} = X_j$, $1 \leq i, j \leq N$ by P_{ij}.

The *joint selection probability*, denoted by $P(Y = X_{(i)}, Y = X_j)$, $1 \leq i, j \leq N$, is defined as the probability that the output equals the ith ranked sample and the jth sample simultaneously, and is given by

$$P(Y = X_{(i)}, Y = X_j) = \frac{P_{ij}}{N!}.$$

Therefore,

$$P(Y = X_{(i)}) = \sum_{j=1}^{N} \frac{P_{ij}}{N!}, \quad i = 1, 2, \ldots, N \tag{4.43}$$

and

$$P(Y = X_j) = \sum_{i=1}^{N} \frac{P_{ij}}{N!}, \quad j = 1, 2, \ldots, N. \tag{4.44}$$

A convenient way of expressing the elements P_{ij} is as the matrix, $\mathbf{P} = \{P_{ij}\}_{N \times N}$. The matrix $\mathbf{P}$ is called the *sample permutation matrix*.

Example 4.10.   Consider a stack filter $S(\cdot)$ defined in the binary domain by the Boolean function $f(x_1, x_2, x_3) = x_1 x_2 + x_3$. Thus, the continuous stack filter is $S(X_1, X_2, X_3) = \mathrm{MAX}\{\mathrm{MIN}\{X_1, X_2\}, X_3\}$. To compute the output for all possible orderings of X_1, X_2, and X_3 we can identify them with the integers $1, 2$ and 3, and explicitly write down the orderings and corresponding outputs.

X_1	X_2	X_3	Y	(i)	j
1	2	3	3	3	3
1	3	2	2	2	3
2	1	3	3	3	3
2	3	1	2	2	1
3	1	2	2	2	3
3	2	1	2	2	2

From this table we get the numbers P_{ij} and dividing by $3! = 6$ gives the joint selection probabilities

$$\left[P\{Y = X_{(i)}, Y = X_j\} \right] = \begin{pmatrix} 0 & 0 & 0 \\ \frac{1}{6} & \frac{1}{6} & \frac{1}{3} \\ 0 & 0 & \frac{1}{3} \end{pmatrix}.$$

Remark 4.10. The rank and sample selection probabilities are in fact scaled row and column sums of the sample permutation matrix because the ith rank selection probability $P(Y = X_{(i)})$, $1 \le i \le N$ is given by

$$P(Y = X_{(i)}) = (i\text{th row sum of the } \mathbf{P}\text{-matrix})/N!$$

and the jth sample selection probability $P(Y = X_j)$, $1 \le j \le N$ is given by

$$P(Y = X_j) = (j\text{th column sum of the } \mathbf{P}\text{-matrix})/N!.$$

The output distribution of a filter having its rank selection probability vector $\mathbf{r}$ can be expressed with the rank selection probabilities and the output distribution functions of order statistics for continuous i.i.d. inputs in the following way. The output distribution function $\Psi(\cdot)$ and the density function $\psi(\cdot)$ of a stack filter having its rank selection probability vector $\mathbf{r} = (r_1, r_2, \ldots, r_N)$ and continuous i.i.d. inputs are given by

$$\Psi(t) = \sum_{i=1}^{N} r_i F_{(i)}(t) \tag{4.45}$$

and

$$\psi(t) = \sum_{i=1}^{N} r_i f_{(i)}(t),$$

where $F_{(i)}(\cdot)$ and $f_{(i)}(\cdot)$ are the distribution function and the density function, respectively, of the ith order statistics for i.i.d. inputs.

Proof. The output distribution $\Psi(x)$ is given by

$$\Psi(t) = P(Y \le t) = \sum_{i=1}^{N} P(Y \le t, Y = X_{(i)})$$

$$= \sum_{i=1}^{N} P(X_{(i)} \le t, Y = X_{(i)})$$

$$= \sum_{i=1}^{N} P(X_{(i)} \le t) P(Y - X_{(i)}).$$

The last equality holds because the output $Y = X_{(i)}$ does not depend on the value of $X_{(i)}$, but only on the order of samples, and the value of $X_{(i)}$ does not depend on whether $X_{(i)}$ is the output or not. Thus,

$$\Psi(t) = \sum_{i=1}^{N} P(Y = X_{(i)}) F_{(i)}(t)$$

holds, and by differentiating with respect to t we obtain $\psi(t)$. $\square$

The condition that the output rank depends only on the permutation that orders the values in the window and not on their actual values is crucial for the expansion (4.45) to hold. This can be seen from the following counterexample where this assumption is not valid.

Example 4.11. Let the random variables X_1 and X_2 be identically distributed and independent with a Gaussian distribution. Define a filter by specifying its output

$$Y(X_1, X_2) = \begin{cases} X_{(1)}, & \text{if } X_{(2)} > 0, \\ X_{(2)}, & \text{if } X_{(2)} \le 0. \end{cases}$$

Because the joint distribution is circularly symmetric the probability of the event $\{X_{(1)} \le 0, Y = X_{(1)}\}$ equals the probability of the event $\{X_{(1)} \le 0, X_{(2)} \ge 0\}$. Because the ordering procedure sends any point below the line $x_1 = x_2$ to its mirror image with respect to this line the probability of this event is twice the probability that the point falls into the second quadrant and is thus $1/2$. On the other hand, similar reasoning shows that $P(X_{(1)} \le 0) = 3/4$ and $P(Y = X_{(1)}) = 3/4$. Thus,

$$P(X_{(1)} \le 0, Y = X_{(1)}) \ne P(X_{(1)} \le 0)P(Y = X_{(1)}).$$

In Section 4.2.1 we derived expressions for the output distribution and density of a stack filter in terms of the coefficients A_i and the distribution and density of the i.i.d. inputs. We now have two expressions for the output distribution, the first as a linear combination of terms of the type $(1 - \Phi(t))^i \Phi(t)^{N-i}$ and the second as a linear combination of the terms $P(X_{(i)} \le t)$. By (4.32) $P(X_{(i)} \le t)$ themselves are linear combinations of terms of type $(1 - \Phi(t))^i \Phi(t)^{N-i}$. It is easy to see that the functions $(1 - \Phi(t))^i \Phi(t)^{N-i}$ are linearly independent.

Thus the coefficients must be equal implying the following linear relation between A_i and r_i. The coefficients A_i of a stack filter $S(\cdot)$ of window size N and the rank selection vector $\mathbf{r} = (r_1, r_2, \ldots, r_N)$ satisfy (Exercise)

$$r_j = \frac{A_{N-j}}{\binom{N}{j}} - \frac{A_{N-j+1}}{\binom{N}{j-1}}, \quad j = 1, 2, \ldots, N, \tag{4.46}$$

or equivalently

$$A_{i+1} = \binom{N}{i+1}\left[\frac{A_i}{\binom{N}{i}} - r_{N-i}\right] = \frac{N-i}{i+1}A_i - \binom{N}{i+1}r_{N-i}, \quad i = 0, 1, \ldots, N-1.$$

Because A_i and r_i are necessarily nonnegative these relations imply

$$A_{j+k} = \binom{N}{j+k}\left[\frac{A_j}{\binom{N}{j}} - \sum_{i=N-j-k+1}^{N-j} r_i\right], \quad 1 \le k \le N-j,$$

$$\binom{N}{j}\sum_{i=N-j-k+1}^{N-j} r_i \le A_j, \quad 1 \le k \le N-j,$$

$$A_{i+1} \le \frac{N-i}{i+1}A_i, \quad i = 0, 1, \ldots, N-1, \tag{4.47}$$

and

$$A_{i+1} \le A_i, \quad \text{when} \quad i \ge \frac{N-1}{2}.$$

It is important to notice that these relations do not depend on the particular stack filter in question. The rank selection probabilities give an intuitively appealing way of constraining a stack filter. For instance, a certain amount of robustness is guaranteed if we require that the rank selection probabilities are zero for a number of the largest and smallest indices. This will give a stack filter that is "trimmed" in the same way as an L-filter with the coefficients corresponding to a number of the largest and smallest coefficients equal to zero. Constraints on the rank selection probabilities translate immediately into constraints on A_i because of the above relations. For instance, if it is required that

$$r_1 = r_2 = \cdots = r_k = 0,$$

then

$$A_{N-1} = A_{N-2} = \cdots = A_{N-k} = 0,$$

and if it is required that

$$r_k = r_{k+1} = \cdots = r_N = 0,$$

then

$$A_j = \binom{N}{j}, \quad j = 1, 2, \ldots, N - k + 1.$$

4.2.5   Optimization of Stack Filters

The above relations make it possible to optimize stack filters in the mean square sense without performing a full search over all stack filters, which would also be impossible except for small window sizes because of the very large number of stack filters. For instance, it is easy to see (Exercise) that for the window size N the number of different stack filters is greater than $2^{2^N/N}$. The optimization consists of finding a solution of the integer linear programming task

$$\text{minimize} \quad \sum_{i=0}^{N-1} A_i M(\Phi, 2, N, i) \tag{4.48}$$

under the constraints

$$c_{11} \leq A_1 \leq c_{12}$$
$$c_{21} \leq A_2 \leq c_{22}$$
$$\vdots$$
$$c_{N-1,1} \leq A_{N-1} \leq c_{N-1,2},$$

and then determining a stack filter with the above coefficients A_i if it exists. It must be emphasized that usually there is no guarantee that such a stack filter exists. However, once we have the target coefficients A_i the search for the optimal stack filter is simpler. We can also take a stack filter that has coefficients A_i close to the solution of the optimization problem (4.48) and then check if its properties are satisfactory.

In the above, we considered the rank selection probabilities and used them to constrain the filter in the statistical sense. In many image processing problems it is not enough to know the "average" behavior of the filter, but we need to be sure that it will handle certain signal segments in a prescribed way. This can be achieved using so called *structural* *constraints*. The goal of the structural constraints is to preserve some desired signal details, e.g., pulses in 1-D signals, lines in images, and to remove undesired signal patterns. The structural constraints consist of a list of different structures to be preserved, deleted or modified. Since stack filters obey the threshold decomposition, the structural constraints need only to be considered in the context of binary signals. That is, they can be specified by a set of binary vectors and their outputs. The binary vectors are divided into two subsets, *type 1* constraints and *type 0* constraints.

A binary vector which is specified by the structural constraints is called a *type 1* constraint if its output is one; otherwise it is called a *type 0* constraint. Denote the set of all *type 1* constraints by $\Gamma_1 = \{x_1, x_2, \ldots, x_p\}$ and the set of all *type 0* constraints by $\Gamma_0 = \{y_1, y_2, \ldots, y_q\}$.

Structural constraints induce two new constraints for the coefficients A_i: Let the number of vectors $x \in \Gamma_1$ with $w_H(x) = i$ be $\gamma_i^{(1)}$ for all $1 \leq i \leq N - 1$. Then

$$A_i \leq \binom{N}{i} - \gamma_i^{(1)}, \quad 1 \leq i \leq N - 1.$$

Let the number of vectors $x \in \Gamma_0$ with $w_H(x) = i$ be $\gamma_i^{(0)}$ for all $1 \leq i \leq N - 1$. Then

$$A_i \geq \gamma_i^{(0)}, \quad 1 \leq i \leq N - 1.$$

Example 4.12. Finding the coefficients A_i of the stack filters of window length $N = 5$ which, for white Gaussian N(0,1) noise, minimize the second order central output moment under given constraints :

- The output is never the largest or the smallest value in the window;

- The probability that the filter outputs the median of the values in the window is at most 0.5.

It is straightforward to show (Exercise) that the constraints are equivalent to the following inequalities:

$$A_4 = 0, \quad A_1 = 5 \tag{4.49}$$

and

$$A_3 \geq \frac{3}{3} A_2 - \frac{\binom{5}{2}}{2} = A_2 - 5. \tag{4.50}$$

Now, using formula (4.40) and inequalities (4.49), (4.50) and (4.47) we see that the optimal coefficients A_i are found by minimizing the expression

$$(1, 5, A_2, A_3, 0) \begin{pmatrix} 1.80002 \\ -0.24869 \\ -0.02697 \\ 0.02697 \\ 0.24869 \end{pmatrix} - \left[(1, 5, A_2, A_3, 0) \begin{pmatrix} 1.16296 \\ -0.13359 \\ -0.04950 \\ -0.04950 \\ -0.13359 \end{pmatrix} \right]^2$$

$$= 0.55657 + 0.02697(A_3 - A_2) - (0.49501 - 0.04950(A_2 + A_3))^2$$

subject to

$$0 \leq A_2 \leq 10, \quad A_2 - 5 \leq A_3 \leq A_2.$$

The coefficients are computed from the recurrence (4.41). Because in this example we do not require the filter to be self-dual, the first moment appears also in the above minimization. By using nonlinear programming the solutions are found to be

$$\mathbf{a}_1 = (1, 5, 10, 10, 0) \quad \text{and} \quad \mathbf{a}_2 = (1, 5, 0, 0, 0), .$$

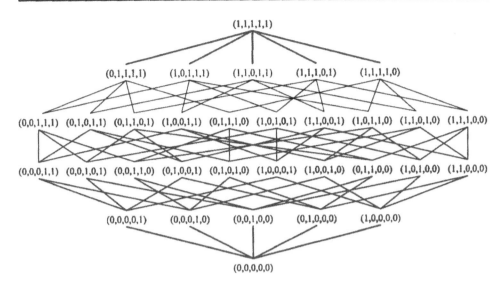

Figure 4.4. Lattice B_5.

which corresponds to the rank selection probability vectors

$$\mathbf{r}_1 = (0, 1, 0, 0, 0), \quad \text{and} \quad \mathbf{r}_2 = (0, 0, 0, 1, 0).$$

Thus, the optimal filters are simply the 2nd and 4th order statistic filters. If we further require that the filter must be self-dual the solution is given by a weighted median filter (Exercise).

4.2.6 Stack Filter Optimization in the Boolean Lattice

In the previous algorithm the optimal A_is were found and then the stack filter having these A_is is searched. As it was mentioned nothing guarantees that it will exist. Now, we present a highly related idea, which can be used to find the optimal filter, not only the A_is under constraints and which sheds more light on the methods in the previous section. The idea is best described in the lattice framework.

We recall some basic lattice terminology [94]. A *partially ordered set* (poset) is a set with a binary relation which is reflexive, antisymmetric and transitive. For example, $B_N = \{\mathbf{x} = (x_1, x_2, \ldots x_N) : x_i \in \{0, 1\}\}$ and the relation $\leq$

$$\mathbf{x} \leq \mathbf{y} \text{ if and only if } x_n \leq y_n \text{ for all } n.$$

is a poset. A *lattice* is a poset L for which every pair of elements has a least upper bound and a greatest lower bound [94]. For example, B_N is a lattice (See Figure 4.4). Let L be a lattice. A *lower ideal* of L is a subset P of L such that if $\mathbf{x} \in P$ and $\mathbf{y} \leq \mathbf{x}$, then $\mathbf{y} \in P$. An *upper ideal* of L is a subset Q of L such that if $\mathbf{x} \in Q$ and $\mathbf{y} \geq \mathbf{x}$, then $\mathbf{y} \in Q$. Let $\mathbf{y}_1, \mathbf{y}_2, \ldots, \mathbf{y}_k \in L$, then $P = \{\mathbf{x} \in L : \exists \mathbf{y}_i, i = 1, 2, \ldots, k : \mathbf{x} \leq \mathbf{y}_i\}$ is called the *lower ideal generated by* $\mathbf{y}_1, \mathbf{y}_2, \ldots, \mathbf{y}_k$, and similarly, $Q = \{\mathbf{x} \in L : \exists \mathbf{y}_i, i = 1, 2, \ldots, k : \mathbf{x} \geq \mathbf{y}_i\}$ is called the *upper ideal generated by* $\mathbf{y}_1, \mathbf{y}_2, \ldots, \mathbf{y}_k$. Let $A \subseteq L$, then $\mathbf{x} \in L$ is called a *maximal element* of A, if $\mathbf{y} \in A, \mathbf{y} \geq \mathbf{x}$, implies $\mathbf{x} = \mathbf{y}$. A *minimal element* of A is defined in a similar manner. It is clear that an upper ideal is generated by its minimal elements and a lower ideal is generated by its maximal elements.

Earlier we noticed that if in the optimization there are no constraints, then the optimal filter has $A_i = \binom{N}{i}$ for $1 \leq i \leq (N-1)/2$ and $A_i = 0$ for $(N+1)/2 \leq i \leq N-1$. This indicates that all the elements of B_N with Hamming weight less than $N/2$ are mapped to 0 and the other elements are mapped to 1 by the filter. The resulting filter is the median filter. When there are constraints, for some vectors $\mathbf{x} \in B_N$, $w_H(\mathbf{x}) \leq N/2$, it holds $f(\mathbf{x}) = 1$ and because of the self-duality, $f(\overline{\mathbf{x}}) = 0$, where $f(\cdot)$ is the positive Boolean function defining the optimal self-dual stack filter. Let Ω_1 be the upper ideal generated by these vectors $\mathbf{x}$ and Ω_0 be the lower ideal generated by the vectors $\overline{\mathbf{x}}$, that is,

$$\Omega_1 = \{\mathbf{x} \in B_N \ : \ w_H(\mathbf{x}) \leq N/2\} \cap f^{-1}(1)$$

and

$$\Omega_0 = \{\mathbf{x} \in B_N \ : \ w_H(\mathbf{x}) \geq N/2\} \cap f^{-1}(0).$$

Thus, $\Omega_1 \subseteq f^{-1}(1)$ and $\Omega_0 \subseteq f^{-1}(0)$. In order that there exists a positive Boolean function $f(\cdot)$ (or a stack filter) with Ω_0 and Ω_1, it must hold $\Omega_0 \cap \Omega_1 = \emptyset$. If the constraints are such that it is possible to form sets Ω_0 and Ω_1 in such a way that $\Omega_0 \cap \Omega_1 = \emptyset$ the constraints are feasible.

Furthermore, we denote $\Omega_{\mathrm{med}} = \{\mathbf{x} \in B_N \ : \ w_H(\mathbf{x}) \geq N/2\}$. Then the filter $S_f(\cdot)$ can be described by

$$f^{-1}(1) = (\Omega_1 \cup \Omega_{\mathrm{med}}) \cap (B_N \setminus \Omega_0). \tag{4.51}$$

Equation (4.51) expressed as a Boolean function yields

$$f(\mathbf{x}) = \Big(\sum_{i=1}^{p} f_i(\mathbf{x}) + f_{\mathrm{med}}(\mathbf{x})\Big) \prod_{i=1}^{p} f_i^D(\mathbf{x}), \tag{4.52}$$

where $f_{\mathrm{med}}(\cdot)$ is the Boolean function of the N point standard median filter, $f_i(\cdot)$ is the elementary conjunction of an element of Ω_1, and p is the number of elements in the set Ω_1.

We can reduce (4.52) into a more simplified form, as we need only to sum over the minimal elements of Ω_1 instead of summing over all members of Ω_1. As is easy to see (Exercise), the complements of the minimal elements of Ω_1 are the maximal elements of Ω_0.

Remark 4.11.   We know that any stack filter can be implemented for multi-level signals by replacing AND and OR with MIN and MAX operations, respectively. This implies the following useful result.

For real-valued signals, the optimal stack filter defined by the positive Boolean function given in (4.52) is a composition of the median filter and a set of maximum and minimum filters:

$$S_f(\mathbf{X}) = \mathrm{MIN}\{\mathrm{MAX}\{\mathrm{MED}\{\mathbf{X}\}, S_1(\mathbf{X})\}, S_2(\mathbf{X})\},$$

where

$$S_1(\mathbf{x}) = \mathrm{MAX}\{S_{f_1}(\mathbf{x}), S_{f_2}(\mathbf{x}), \ldots, S_{f_p}(\mathbf{x})\},$$

$$S_2(\mathbf{x}) = \mathrm{MIN}\{S_{f_1}^D(\mathbf{x}), S_{f_2}^D(\mathbf{x}), \ldots, S_{f_p}^D(\mathbf{x})\},$$

and $S_{f_i}(\mathbf{x})$, $i = 1, 2, \ldots, p$, are stack filters for real-valued signals corresponding to the conjuctions $f_i(\mathbf{x})$. Likewise, $S_{f_i}^D(\mathbf{x})$, $i = 1, 2, \ldots, p$ correspond to the duals of the conjuctions $f_i(\mathbf{x})$.

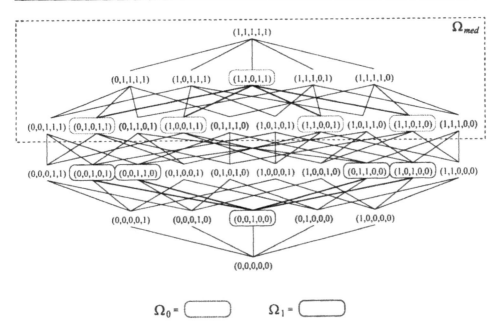

$$\Omega_0 = \boxed{} \qquad \Omega_1 = \boxed{}$$

Figure 4.5. Lattice B_5 with structural constraints shown. The constraints and their consequences are shown by the boxes with rounded corners. The set that is mapped to 1 by the Boolean function defining the median filter is shown by the large box with dotted boundary.

The idea of our algorithm is to choose the minimal elements of Ω_1 in such a way that the constraints are satisfied and that the cardinalities of the sets Ω_1 and Ω_0 are kept as small as possible. This aim is achieved by adding new minimal elements only to the levels (by ith level we mean the vectors $\mathbf{x}$ with $w_H(\mathbf{x}) = i$) where they are necessary and as few of them as possible. By choosing as few elements as possible we leave the maximum amount of freedom for the next levels, because we have maximized the amount of vectors which may or may not belong to the sets Ω_1 and Ω_0 in the next levels. This freedom makes it usually easier to satisfy given constraints.

Example 4.13. Assume that we have a structural constraint $\Gamma_1 = \{(0,0,1,0,0)\}$ and that we wish to find the optimal self-dual stack filter under this constraint. As $(0,0,1,0,0) \in \Gamma_1$ it must hold that $f(0,0,1,0,0) = 1$, thus, $(0,0,1,0,0) \in \Omega_1$. The self-duality requirement implies that $f(1,1,0,1,1) = 0$, thus $(1,1,0,1,1) \in \Omega_0$. As we are considering positive Boolean functions satisfying $f(0,0,1,0,0) = 1$, it also should hold that $f(\mathbf{x}) = 1$ for all $\mathbf{x} \geq (0,0,1,0,0)$. This means that also

$$\{(0,0,1,0,1),(0,0,1,1,0),(0,1,1,0,0),(1,0,1,0,0)\} \subset \Omega_1$$

which further implies by the self-duality that

$$\{(1,1,0,1,0),(1,1,0,0,1),(1,0,0,1,1),(0,1,0,1,1)\} \subset \Omega_0.$$

See Figure 4.5. Now it is easy to see that $A_0 = 1, A_1 = 4, A_2 = 6, A_3 = 4, A_4 = 1$, and we are fairly far away from the median which is optimal in the case without constraints. Can we now make any change to find a better self-dual stack filter satisfying the constraints?

The only changes that we could do would be to change some vectors of Hamming weight two that are now mapped to 0 to be mapped to 1 and to update the sets Ω_1 and Ω_0 remembering the self-duality requirement. In this case, it is easy to see from Figure 4.5 that we can make no changes without violating the positivity of the Boolean function. Even if it had been possible these changes would have resulted in a decrease in the values of A_is for $1 \leq i \leq (N-1)/2$ and increase in the values of A_is for $(N+1)/2 \leq i \leq N-1$. This would have caused an increase in the output variance and thus the resulting filter would not be optimal.

From the previous considerations we can derive the following simple optimization method for the cases with only structural constraints.

Remark 4.12.   Assume that there are only structural constraints. Then the Boolean function corresponding to the optimal self-dual stack filter equals

$$f(\mathbf{x}) = (\sum_{i=1}^{p} f_i(\mathbf{x}) + f_{\text{med}}(\mathbf{x})) \prod_{j=1}^{p} f_j^D(\mathbf{x}), \qquad (4.53)$$

where $f_{\text{med}}(\cdot)$ is the Boolean function of the N point standard median filter, $f_i(\cdot)$ is the elementary conjunction of an element of Γ_1, and p is the number of elements in the set Γ_1.

4.3   Multistage and Hybrid Filters

Many nonlinear filters in Chapter 3 could be interpreted as stack filters where the inputs are outputs of certain subfilters. For example a multistage median filter defined in Section 3.7 is a cascade of median filters. We know that each of the median filters is a stack filter and thus the overall filter is a cascade of stack filters. A stack filter could be equivalently characterized by the defining positive Boolean function which means that the multistage median filters is a stack filter where the defining positive Boolean function is the compound Boolean function of the Boolean functions that define the medians on each stage.

Example 4.14.   Consider the simplest possible case where we have a one-dimensional input signal X_i and the output signal Y_i is defined by

$$Y_i = \text{MED}\{\text{MED}\{X_{i-2}, X_{i-1}, X_i\}, \text{MED}\{X_{i-1}, X_i, X_{i+1}\}, \text{MED}\{X_i, X_{i+1}, X_{i+2}\}\}.$$

The corresponding expression with Boolean functions is

$$y_i = f(f(X_{i-2}, X_{i-1}, X_i), f(X_{i-1}, X_i, X_{i+1}), f(X_i, X_{i+1}, X_{i+2}))$$

where $f(u, v, w) = uv + uw + vw$. A straightforward calculation shows that the composite Boolean function that corresponds to the multistage median simplifies to

$$y_i = x_{i-2}x_{i-1}x_{i+1} + x_{i-2}x_ix_{i+2} + x_{i-1}x_{i+1}x_{i+2} + x_{i-1}x_i + x_ix_{i+1}.$$

For i.i.d. input the output distribution of this multistage median filter is readily computed using (4.28) (Exercise). It is obvious that this multistage median is equivalent to filtering the signal twice by the median filter of length three. The Boolean function 4.14 is in fact linearly separable and self-dual from which it follows that this multistage median filter is also expressible as a weighted median filter (Exercise).

Linear median hybrid filters that were introduced in Section 3.8 are obtained by changing the subfilters in multistage median filters into linear filters. It follows that the inputs of the second stage are no longer independent and identically distributed even if the original inputs were such. If the subfilter windows do not overlap then the independence is not destroyed, but, depending on the form of the subfilters, their outputs may no longer be identically distributed. The more general formula (4.27) for the output distribution still applies.

The simplest linear median hybrid filter introduced in Section 3.8 uses a three point median over three nonoverlapping linear filters; the forward predictor, the center point (identity filter), and the backward predictor. In the following we analyze the output distribution of this filter for independent Gaussian inputs.

The basic linear median hybrid was defined in Section 3.8 by

$$\text{FMH}(X_1, X_2, \ldots, X_N) = \text{MED}\{F_1(X_1, X_2, \ldots, X_k), X_{k+1}, F_2(X_{k+2}, X_{k+3}, \ldots, X_N)\},$$

where $F_1(X_1, X_2, \ldots, X_k)$ and $F_2(X_{k+2}, X_{k+3}, \ldots, X_N)$ are mean filters and thus linear filters. Because the output of a linear filter for Gaussian inputs is also Gaussian and the median operates on three inputs the overall output distribution stays simple enough to permit computation of the output variance.

Let F, H and G be the distribution functions of the outputs of F_1, F_2 and the center sample respectively and f, g and h be the corresponding density functions. Using (4.27) one easily sees (Exercise) that the output density function of the whole filter is

$$m = f(G + H) + g(H + F) + h(F + G) - 2(fGH + gHF + hFG). \tag{4.54}$$

Using this expression the output variance can be calculated once the output distributions of the subfilters have been obtained. If the input is white Gaussian noise this is particularly simple and in the following we shall derive a formula for the output variance in this case. The derivation is based on the following fact:

Suppose that U_i, $i = 1, 2, 3$ are independent normally distributed random variables with density and distribution functions

$$f_i(x) = \frac{1}{\sqrt{2\pi}\sigma_i} e^{-\frac{x^2}{2\sigma_i^2}}, \quad F_i(x) = \int_{-\infty}^{x} f_i(t)dt.$$

Then the variance of the random variable $Y = \text{MED}\{U_1, U_2, U_3\}$ is

$$Var\{U\} = \frac{1}{\pi}(\sum_{i=1}^{3} \sigma_i^2 \arctan \frac{p}{\sigma_i^2} - p), \tag{4.55}$$

where $p = \sqrt{\sigma_1^2\sigma_2^2 + \sigma_2^2\sigma_3^2 + \sigma_3^2\sigma_1^2}$.

Because the expectation of Y is obviously zero the variance is, in principle, obtained by simply integrating $x^2 m(x)$. To illustrate the technical difficulties that we often encounter when we are dealing with densities that differ from standard tabulated distributions we give the main steps of the derivation [10, 28]. Thus

$$Var\{Y\} = \int_{-\infty}^{\infty} x^2 m(x)dx, \tag{4.56}$$

where $m(x)$ is the density of Y. By (4.54) this reduces to integrals of the form

$$I_1 = \int_{-\infty}^{\infty} x^2 f_1(x)F_2(x)dx$$

and

$$I_2 = \int_{-\infty}^{\infty} x^2 f_1(x) F_2(x) F_3(x) dx.$$

Now, integration by parts gives

$$I_1 = \int_{-\infty}^{\infty} \sigma_1^2 f_1(x) F_2(x) dx - \int_{-\infty}^{\infty} \sigma_1^2 x f_1(x) f_2(x) dx$$

$$= \sigma_1^2 \int_{-\infty}^{\infty} dx \int_{-\infty}^{x} f_1(x) f_2(t) dt = \frac{\sigma_1^2}{2}.$$

To calculate I_2 we write, after integrating once by parts,

$$I_2 = \sigma_1^2 \{ \int_{-\infty}^{\infty} f_1(x) F_2(x) F_3(x) dx + \int_{-\infty}^{\infty} x f_1(x) f_2(x) F_3(x) dx + \int_{-\infty}^{\infty} x f_1(x) f_3(x) F_2(x) dx \}.$$

The last integral above can be written as

$$\int_{-\infty}^{\infty} x f_1(x) f_2(x) F_3(x) dx = -\frac{\sigma_2^2}{\sigma_1^2} \int_{-\infty}^{\infty} x f_1(x) f_2(x) F_3(x) dx + \sigma_2^2 \int_{-\infty}^{\infty} f_1(x) f_2(x) f_3(x) dx$$

giving

$$\int_{-\infty}^{\infty} x f_1(x) f_2(x) F_3(x) dx = \frac{\sigma_1^2 \sigma_2^2}{2\pi p (\sigma_1^2 + \sigma_2^2)},$$

where we have written

$$p = \sqrt{\sigma_1^2 \sigma_2^2 + \sigma_2^2 \sigma_3^2 + \sigma_3^2 \sigma_1^2}.$$

Similarly,

$$\int_{-\infty}^{\infty} x f_1(x) f_3(x) F_2(x) dx = \frac{\sigma_1^2 \sigma_3^2}{2\pi p (\sigma_1^2 + \sigma_3^2)}.$$

Further, we can write

$$\int_{-\infty}^{\infty} f_1(x) F_2(x) F_3(x) dx = \int_{-\infty}^{\infty} \{ f_1(x) \int_{-\infty}^{x} f_2(y) dy \int_{-\infty}^{x} f_3(z) dz \} dx,$$

which by transformation

$$\begin{cases} u & = x/\sigma_1 \\ v & = y/\sigma_2 \\ w & = z/\sigma_3 \end{cases}$$

reduces to

$$(\frac{1}{2\pi})^{3/2} \int_{-\infty}^{\infty} du \int_{-\infty}^{\frac{\sigma_1 u}{\sigma_2}} dv \int_{-\infty}^{\frac{\sigma_1 u}{\sigma_3}} e^{-\frac{1}{2}(u^2+v^2+w^2)} dw = \frac{1}{2\pi} (\pi - \arccos \frac{\sigma_1^2}{\sqrt{(\sigma_1^2 + \sigma_2^2)(\sigma_1^2 + \sigma_3^2)}}).$$

Combining the above computations we finally get

$$I_2 = \sigma_1^2 \{ \frac{1}{2} - \frac{1}{2\pi} \arccos \frac{\sigma_1^2}{\sqrt{(\sigma_1^2 + \sigma_2^2)(\sigma_1^2 + \sigma_3^2)}} + \frac{\sigma_1^2(p^2 + \sigma_2^2 \sigma_3^2)}{2\pi p (\sigma_1^2 + \sigma_2^2)(\sigma_1^2 + \sigma_3^2)} \}.$$

Using the expressions for I_1, I_2 and I_3 in the integrals that are obtained from substituting (4.54) in (4.56) we finally get (4.55). Using the above formulas we can now derive output variances for linear median hybrid filters.

Example 4.15. Consider the basic linear median hybrid filter. Let the input signal be white Gaussian noise with $\sigma^2 = 1$. The outputs of the subfilters are independent Gaussian random variables with variances $\sigma_1^2 = \frac{1}{k}$, $\sigma_2^2 = 1$ and $\sigma_3^2 = \frac{1}{k}$. Thus the output variance is

$$\sigma_o^2 = \frac{1}{\pi}(\frac{2}{k}\arctan\sqrt{2k+1} + \arctan\frac{\sqrt{2k+1}}{k} - \frac{\sqrt{2k+1}}{k}),$$

which for large k can be approximated as

$$\sigma_o^2 = \frac{1}{k} + O(k^{-3/2}).$$

This indicates that for large k on constant plus noise sections of the signal the noise attenuation corresponds to that of a single averaging filter of half length.

Consider the filtering of edges with added white noise, i.e., signals of the form

$$X_n = S_n + N_n,$$

where

$$S_n = \begin{cases} 0, & \text{for } n < 0, \\ h, & \text{for } n \geq 0, \end{cases}$$

and N_n are i.i.d. random variables.

If N_n is white Gaussian noise, it is possible to express the bias and the variance near the edge as a sum of integrals of products of normal densities and distributions. However, the expressions are so complicated that it is simpler to use direct numerical integration.

If the height of the edge h is large then just before the edge the bias and variance are the expectation and variance of $U_n = \text{MAX}\{X_{-1}, F_1(X_{-k-1}, X_{-k}, \ldots, X_{-2})\}$, where X_{-1} and $F_1(X_{-k-1}, X_{-k}, \ldots, X_{-2})$ are Gaussian, have zero mean, and variances 1 and $\frac{1}{k}$ respectively. Thus

$$\text{bias}_{max} = \sqrt{\frac{k+1}{2\pi k}}$$

and

$$\text{variance}_{max} = \frac{(\pi - 1)(k+1)}{2\pi k}.$$

With a moving average filter of length $2k + 1$ the variance stays equal to $\frac{1}{2k+1}$ all the time but at the edge the bias error is as large as $\frac{hk}{2k+1}$.

Similar methods apply to many other filter classes as well. For instance, the in place growing structures can be forced into the stack filter formalism. Usually the main problem is that because of the complicated subfilter structure the independence is lost and the output distribution must be computed directly from the definition. This is possible only for small sizes of the filter window.

4.4 Discussion

We have studied the statistical properties of three classes of nonlinear filters, L-filters, stack filters, and linear median hybrid filters. We hope that the reader has learned at least the following lessons from this very brief and necessarily superficial treatment of the properties of nonlinear filters. The first lesson is that for certain filter classes the theory is quite advanced and, for instance, there are sophisticated methods of designing filters for particular applications. Unfortunately, the second lesson is that the theory is still far from being sufficiently mature to permit the development of standard procedures that would allow a practicing engineer to routinely design nonlinear filters. The third lesson might be that, even though the theory is not mature, it is already quite deep and the behavior of many filters is well understood. This means that any significant advance in practical filtering methods is hardly possible without using the sophisticated tools of estimation theory and random processes. A typical example of possible future breakthroughs is stack filters. Restricted optimization procedures can now be performed under the assumption of white noise, and methods that would allow us to handle the case of dependent noise would immediately lead to great improvements in actual filtering techniques.

WE SHALL NOT CEASE FROM EXPLORATION, AND THE END OF ALL OUR EXPLORING SHALL BE TO ARRIVE WHERE WE STARTED AND KNOW THE PLACE FOR THE FIRST TIME.
T. S. Eliot, 1888–1965

5

EXERCISES

THIS STUDY YOU OUGHT TO PRACTICE FROM MORNING TO EVENING,
BEGINNING FROM THE SMALLEST THINGS...
Epictetus, "Discourses IV", Circa 60 AD

I HEAR, AND I FORGET;
I SEE, AND I REMEMBER;
I DO, AND I UNDERSTAND.
Ancient Chinese philosophy

Chapter 1: NONLINEAR SIGNAL PROCESSING.

1. Verify whether or not $f(x) = x^2$ and $g(x) = nx$ are linear.

2. (For those who are familiar with linear filtering.) Analyze the frequency domain behavior of the three signal segments used in Section 1.3.1.

3. What kind of a linear filter do we have if its impulse response is a zero-valued sequence?

4. Discuss how we can filter the signal areas where the moving window does not fit completely inside the signal.

5. Find an algorithm for finding the maximum of N values $X_1, X_2, \ldots, X_N$ by using $N - 1$ comparisons.

6. (For those who are familiar with linear filtering.) Derive the frequency response of a mean filter with window length N. Conclude that the only signal with finite energy that can pass the mean filter unaltered is the zero signal.

7. Write down the difference equation that a signal invariant (See Definition 5.1) to the mean filter of length 3 satisfies. Generalize to arbitrary window size N. Characterize the signals that are invariant to the mean filter in the range $a, a + 1, \ldots, b, (b > a)$. What happens to the energy of this signal if $a \to -\infty$ and $b \to \infty$?

Chapter 2: STATISTICAL PRELIMINARIES.

1. Consider the Cauchy distribution

$$f_X(\xi) = \frac{1}{\pi(1 + \xi^2)}, \quad -\infty < \xi < \infty.$$

(a) Show that all its moments of positive order are nonexistent.

(b) Find the values of ξ_0 for which $F_X(\xi_0) = 0.7, 0.8$, and 0.9.

(c) Show that all the moments of the truncated distribution

$$\text{constant} \times \frac{1}{(1 + \xi^2)}, \quad -\xi_0 < \xi < \xi_0$$

exist and are finite. Give the formula of the constant as a function of ξ_0.

(d) Show also that the approximation

$$P(X > \xi_0) \approx \frac{1}{\pi \xi_0}$$

holds for large ξ_0.

2. Show that the random variable in (2.1) does not have the median.

3. Consider a random variable defined on $[0, \infty)$ by its distribution function

$$F_X(t) = 1 - e^{-t},$$

the exponential distribution. Let Y be the random variable defined by $Y = X^2$. Compute the expectation $E\{Y\}$ of Y (a) by first deriving the distribution function and density of Y, (b) by using the formula for the expectation of a function of a random variable. Notice that one has to perform almost the same actual calculations in both cases.

4. Consider a random variable X and assume that $M(\theta) = E\{e^{\theta X}\}$ exists. The function M is called the *moment generating function* of X because

$$\frac{d^n}{d\theta^n} M(\theta)|_{\theta=0} = E\{X^n\}.$$

Justify this by (a) performing differentiations under the integral sign in the definition of $E\{e^{\theta X}\}$, and (b) first expanding $e^{\theta X}$ as a power series of θ and then using the linearity of the expectation.

5. Use the facts that

$$\int_{\xi_0}^{\infty} e^{-\frac{\xi^2}{2}}\,d\xi = \frac{1}{\xi_0}e^{-\frac{\xi_0^2}{2}} - \int_{\xi_0}^{\infty}\frac{1}{\xi^2}e^{-\frac{\xi^2}{2}}\,d\xi$$

and that for $\xi > 0$

$$\int_{\xi_0}^{\infty}\frac{1}{\xi^2}e^{-\frac{\xi^2}{2}}\,d\xi \le \frac{1}{\xi_0^3}\int_{\xi_0}^{\infty}\xi e^{-\frac{\xi^2}{2}}\,d\xi$$

to prove the approximation in Example 2.5.

6. Show that (2.9) minimizes (2.8).

7. Outline an algorithm for generating normal random number variables by using the formulas in Example 2.15 assuming that you have a random generator generating uniformly $U(0,1)$ distributed data.

8. Let the joint density function of the random variables X and Y be

$$f_{X,Y}(\xi,\eta) = \begin{cases} 1, & \text{if } -1/2 \le \xi \le 1/2,\ -1/2 \le \eta \le 1/2, \\ 0, & \text{otherwise.} \end{cases}$$

Derive the distribution and density functions of X and Y and their joint distribution function. Deduce that X and Y are independent.

9. Let the joint density function of X and Y be

$$f_{X,Y}(\xi,\eta) = \begin{cases} 1, & \text{if } -1/2 \le \xi+\eta \le 1/2,\ -1/2 \le \xi-\eta \le 1/2, \\ 0, & \text{otherwise.} \end{cases}$$

(This density has been obtained by rotating the density of the previous exercise by 45 degrees.) (a) Show that X and Y are not independent. (b) Derive the distribution function and the density of the conditional random variable $X|Y$.

10. Compute the marginal densities in Example 2.10 and verify that the random variables X_1 and X_2 have the variances σ_1^2 and σ_2^2 respectively.

11. Let the random variables X and Y have the joint density function $f_{X,Y}(\xi,\eta)$. Show that the function $T(\cdot)$ that minimizes $E\{(X-T(Y))^2\}$ is the conditional expectation $E\{X|Y\}$. (Hint: Write $f_{X,Y}(\xi,\eta) = f_{X|Y}(\xi|\eta)f_Y(\eta)$ in the integral that defines $E\{(X-T(X))^2\}$ and use the fact that for a random variable Z its mean μ_Z minimizes $E\{(\alpha - Z)^2\}$ as a function of alpha.)

12. Let $X_1, X_2, \ldots, X_N$ and S be random variables that are defined in the same probability space. Show that the coefficients that minimize $E\{(S - (a_1 X_1 + a_2 X_2 + \cdots + a_N X_N))^2\}$ satisfies

$$
\begin{array}{ccccccc}
r_{11}a_1 & + & r_{12}a_2 & + & \cdots & + & r_{1N}a_N & = & r_1 \\
r_{21}a_1 & + & r_{22}a_2 & + & \cdots & + & r_{2N}a_N & = & r_2 \\
& & & & \vdots & & & & \\
r_{N1}a_1 & + & r_{N2}a_2 & + & \cdots & + & r_{NN}a_N & = & r_N,
\end{array}
$$

where $r_{ij} = E\{X_i X_j\}$ form the correlation matrix of the observations and $r_i = E\{SX_i\}$ form the correlation vector between S and the observations.

13. Show that if the linear estimator is required to be location invariant, the weights must satisfy $\sum_{i=1}^{N} a_i = 1$.

14. Show that in Example 2.17, if $\sigma_i^2 = \sigma^2$, $i = 1, 2, \ldots, N$, the a_is minimizing $E\{(S - (a_1 X_1 + \ldots + a_N X_N))^2\}$ are $a_1 = a_2 = \cdots = a_N = \frac{1}{N + \sigma^2/\sigma_S^2}$. Explain in intuitive terms the influence of σ^2 and σ_S^2 on the coefficients.

15. Consider a random variable X with a positive continuous density function $f_X(x)$. Show that the function

$$\rho(T) = \int_{-\infty}^{\infty} |x - T| f_X(x) dx$$

is minimized if T is the median of (the distribution of) X. Use this result to show that the general minimum mean absolute error estimator is the conditional median (2.15).

16. Let X_1, X_2, X_3, and S have joint Gaussian distribution

$$f(s, x_1, x_2, x_3) = \frac{1}{(2\pi)^2} \exp\{-\frac{1}{2}\left((s - x_1)^2 + (s - x_2)^2 + (s - x_3)^2\right)\}.$$

Show that the conditional median $\text{MED}\{S|x_1, x_2, x_3\} = \frac{1}{3}(x_1 + x_2 + x_3)$.

17. Give a detailed proof for the fact that the median minimizes $\sum_{i=1}^{N} |x_i - \theta|$.

18. Show that $\hat{\theta} = \frac{1}{N} \sum_{i=1}^{N} X_i$ minimizes $\sum_{i=1}^{N} (X_i - \theta)^2$.

19. Find the maximum likelihood estimator of location for the case of N i.i.d. uniformly distributed samples on the interval (a, b), i.e., samples with the density function

$$f_U(\xi) = \begin{cases} 1/(b-a), & a \le \xi \le b, \\ 0, & \text{otherwise.} \end{cases}$$

20. Find the maximum likelihood estimator of scale parameter θ for the case of N i.i.d. exponentially distributed samples, i.e., samples with the density function

$$f_\theta(\xi) = \frac{1}{\theta} e^{-\frac{\xi}{\theta}}, \quad \xi > 0.$$

21. Find the maximum likelihood estimator of scale for the case of N i.i.d. normally distributed samples.

22. Show that the central moments of a random vector $(X_1, X_2, \ldots, X_k)$ with Dirichlet distribution $D(\alpha_1, \alpha_2, \ldots, \alpha_k; \beta)$ are

$$E\{X_1^{r_1} X_2^{r_2} \cdots X_k^{r_k}\} = \frac{\Gamma(\alpha_1 + r_1) \cdots \Gamma(\alpha_k + r_k)\Gamma(\alpha_1 + \cdots + \alpha_k + \beta)}{\Gamma(\alpha_1 + \cdots + \alpha_k + \beta + r_1 + \cdots + r_k)\Gamma(\alpha_1 + \cdots + \alpha_k)}.$$

Chapter 3: 1001 SOLUTIONS

For every section in this chapter we have some common exercises. Thus, we will not repeat writing down the exercise statements again and again here. For these exercises we need some definitions.

Definition 5.1. A filter $F(\cdot)$ is called *scale invariant* if for a real number a it holds that

$$F(aX_1, aX_2, \ldots, aX_N) = aF(X_1, X_2, \ldots, X_N).$$

A filter $F(\cdot)$ is called *translation invariant* if for a real number b it holds that

$$F(X_1 + b, X_2 + b, \ldots, X_N + b) = F(X_1, X_2, \ldots, X_N) + b.$$

Definition 5.2. The *breakdown point* of a filter is the minimum fraction of outliers that can lead to the situation that the output is also an outlier.

Example 5.1. The breakdown point of the mean filter is $1/N$ as one outlier can carry the mean to another extremely small/large value, i.e., to an outlier. The breakdown point of the median is $(k + 1)/N$ as less than k outliers can be rejected but $k + 1$ same sided outliers will cause the output to be one of them.

Thus, the breakdown point is a global measure of the robustness of filters. High values (close to 0.5) indicate a certain type of robustness.

Definition 5.3. A signal f is a *root* or a *fixed point* or an *invariant* of a filter $F(\cdot)$ if it is invariant under filtering by $F(\cdot)$.

Example 5.2. Consider median filtering with $N = 3$. Figure 5.1 shows how starting from a signal, after two iterations a root signal of the filter is found.

The invariants reveal the deterministic behavior of the filter, and in this respect the set of invariants resembles the passband of a linear frequency selective filter.

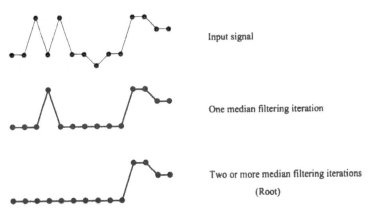

Input signal

One median filtering iteration

Two or more median filtering iterations
(Root)

Figure 5.1. Finding a root of the median filter ($N = 3$).

1. The first common exercise is to verify whether or not the filters discussed in the section are scale (translation) invariant. In some cases some hints are given for the parameters for which these invariance properties might hold.

2. The second common exercise is to find the breakdown points of the discussed filters.

3. Figure 5.2 shows some small regions from our example images. These regions have been shown to be important in a sense that the outputs of different filters often differ essentially in them. Thus, the reader is strongly encouraged to filter these small images with the studied filters in order to obtain a deeper insight as to how the filters behave in very common image structures. The third common exercise will be to filter these images.

4. The fourth common exercise is to estimate the number of

   - additions
   - multiplications
   - compare/swap operations
   - memory cells

   needed to calculate the output at one window position.

5. Verify which of the following signals: step edges, ramp edges, non-increasing/non-decreasing signals, and oscillations ($\{\ldots, a, b, a, b, a, \ldots\}$), are invariants of the filter under consideration. Find also some two-dimensional root signals.

Section 3.1: Trimmed Mean Filters.

1. Study the effect of the parameters r and s on the step response of the (Winsorized) (r, s)-fold trimmed mean filter.

2. Let $\alpha = j/N$, where $0 \leq j \leq N/2$ is an integer. The α-trimmed complementary mean is defined as [14]

$$\text{ComplTrMean}(X_1, X_2, \ldots, X_N; \alpha) = \frac{X_{(1)} + \ldots + X_{(\alpha N)} + X_{(N-\alpha N+1)} + \ldots + X_{(N)}}{2\alpha N}.$$

It is known that this estimator behaves well in thin-tailed distributions like the uniform distribution. What is the impulse and the step response of the α-trimmed complementary mean filter?

Section 3.2: Other Trimmed Mean Filters.

1. Consider a modification of the modified trimmed mean filter where the median in (3.1), or the central pixel X^* in (3.2), is replaced with the mean. What are the practical problems involved so that this definition cannot be used as a filter?

2. Some forms of the trimmed mean filters trim different amounts of samples from different sides of the median. This kind of form can be used for modified trimmed mean filters also. Give a definition of this further modification of the modified trimmed mean filters and describe a situation where this kind of modification could prove useful.

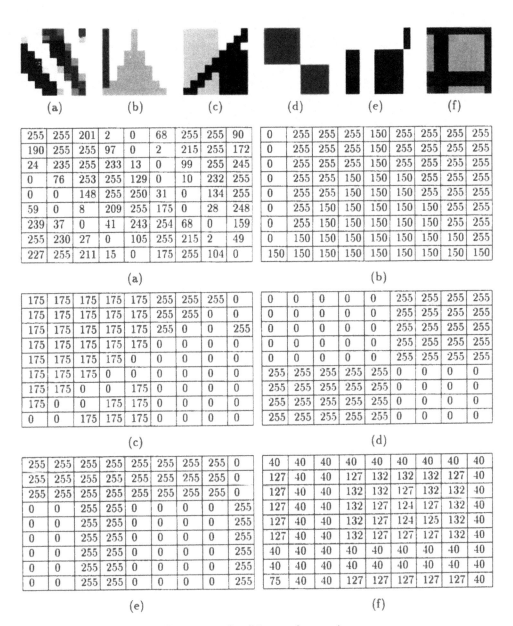

(a)	(b)	(c)	(d)	(e)	(f)

255	255	201	2	0	68	255	255	90
190	255	255	97	0	2	215	255	172
24	235	255	233	13	0	99	255	245
0	76	253	255	129	0	10	232	255
0	0	148	255	250	31	0	134	255
59	0	8	209	255	175	0	28	248
239	37	0	41	243	254	68	0	159
255	230	27	0	105	255	215	2	49
227	255	211	15	0	175	255	104	0

(a)

0	255	255	255	150	255	255	255	255
0	255	255	255	150	255	255	255	255
0	255	255	255	150	255	255	255	255
0	255	255	150	150	150	255	255	255
0	255	255	150	150	150	255	255	255
0	255	150	150	150	150	255	255	255
0	255	150	150	150	150	255	255	255
0	150	150	150	150	150	150	150	255
150	150	150	150	150	150	150	150	150

(b)

175	175	175	175	175	255	255	255	0
175	175	175	175	175	255	255	0	0
175	175	175	175	175	255	0	0	255
175	175	175	175	175	0	0	0	0
175	175	175	175	0	0	0	0	0
175	175	175	0	0	0	0	0	0
175	175	0	0	175	0	0	0	0
175	0	0	175	175	0	0	0	0
0	0	175	175	175	0	0	0	0

(c)

0	0	0	0	0	255	255	255	255
0	0	0	0	0	255	255	255	255
0	0	0	0	0	255	255	255	255
0	0	0	0	0	255	255	255	255
0	0	0	0	0	255	255	255	255
255	255	255	255	255	0	0	0	0
255	255	255	255	255	0	0	0	0
255	255	255	255	255	0	0	0	0
255	255	255	255	255	0	0	0	0

(d)

255	255	255	255	255	255	255	255	0
255	255	255	255	255	255	255	255	0
255	255	255	255	255	255	255	255	0
0	0	255	255	0	0	0	0	255
0	0	255	255	0	0	0	0	255
0	0	255	255	0	0	0	0	255
0	0	255	255	0	0	0	0	255
0	0	255	255	0	0	0	0	255
0	0	255	255	0	0	0	0	255

(e)

40	40	40	40	40	40	40	40	40
127	40	40	127	132	132	132	127	40
127	40	40	132	132	127	132	132	40
127	40	40	132	127	124	127	132	40
127	40	40	132	127	124	125	132	40
127	40	40	132	127	127	127	132	40
40	40	40	40	40	40	40	40	40
40	40	40	40	40	40	40	40	40
75	40	40	127	127	127	127	127	40

(f)

Figure 5.2. Small images for exercises.

Section 3.3: *L*-**Filters.**

1. Consider symmetric *L*-filters with weights satisfying

$$a_i = a_{N+1-i}.$$

Let a and b be constants. Show that the symmetric *L*-filter satisfies:

$$L(aX_1 + b, aX_2 + b, \ldots, aX_N + b; \mathbf{a}) = aL(X_1, X_2, \ldots, X_N; \mathbf{a}) + b\sum_{i=1}^{N} a_i.$$

Why does this hold only for symmetric weights?

2. Show that the median, mean, (r, s)-fold trimmed means, α-trimmed means, and Winsorized means are all special cases of *L*-filters.

3. What could be the reason for the normalization of the weights by $\int_0^1 h(\lambda)d\lambda$ in (3.3)?

4. One generalized ordering is the so-called *median ordering*, where the ordering relation is defined by

$$X_{[i]} \overset{\sim}{\leq} X_{[j]} \text{ if and only if } |X_{[i]} - X_{(k+1)}| \leq |X_{[j]} - X_{(k+1)}|.$$

In other words the samples are ordered by their distance from the median. What kind of weight vector gives now the standard median filter? Where might this ordering be useful? Invent some other generalized orderings.

Section 3.4: *C*-**Filters.**

1. Show how we can obtain FIR filters and *L*-filters as special cases of *C*-filters.

2. One possible approach to solving the gain problem is to normalize the output at every window position by the sum of those N weights used in the output computation. Remembering that the weights can be positive and negative, describe a possible drawback of this approach and how it could be avoided.

3. Show that for every permutation there exists at least one nonzero weight if we can pick an $n \times m$, $n + m = N + 1$, submatrix of $\mathbf{C}$ where all the entries are nonzero. Clearly, this kind of matrix is desirable in many *C*-filtering applications, and now we have an easy way to verify it.

Section 3.5: Weighted Median Filters.

1. Show that the two definitions (3.7) and (3.8) are equivalent.

2. Let a distance matrix $\mathbf{D}$ be defined for the inputs as follows:

$$\mathbf{D} = \begin{pmatrix} d_{11} & d_{12} & \ldots & d_{1N} \\ d_{21} & d_{22} & \ldots & d_{2N} \\ \vdots & \vdots & \ddots & \vdots \\ d_{N1} & d_{N2} & \ldots & d_{NN} \end{pmatrix},$$

where $d_{ij} = |X_i - X_j|$. Let the weights a_i be stored in a weight vector $\mathbf{a} = (a_1, a_2, \ldots, a_N)$ and for the vector $\mathbf{m}^T = \mathbf{D}\mathbf{a}^T$. How can the weighted median be easily found from $\mathbf{m}$?

3. Let $h(\cdot)$ be a monotonous function. Show that

$$h(\text{WeightMed}(X_1, X_2, \ldots, X_N; \mathbf{a})) = \text{WeightMed}(h(X_1), h(X_2), \ldots, h(X_N); \mathbf{a}).$$

This invariance property indicates that the weighted median filter is not disturbed by (even nonlinear) scaling but also that the weighted median filter cannot utilize the possibly useful information in the magnitude differences of samples.

4. What is the effect on the output of a weighted median filter if all the weights are multiplied by a positive constant?

5. When will the center weighted median filter with the center weight $N - 2$ change the value of the signal under filtering?

6. Describe possible problems in the weighted median filter implementation if negative weights can be used.

Section 3.6: Ranked-Order and Weighted Order Statistic Filters.

1. Describe how the rth ranked-order filter transforms step edges for different values of r.

2. Show that the weighted order statistic filters completely remove an impulse if for every weight a_i it holds that $a_i < \min\{r, 1 - r + \sum_{j=1}^{N} a_j\}$.

Section 3.7: Multistage Median Filters.

1. What is the output of the separable median filter operating first on columns and then on rows when applied to the same image used in Example 3.11?

2. Assume that we use for finding the median of N samples an algorithm that has complexity $\approx \alpha N \log N$. For what values of N is the separable median filter faster than the median filter of the same window size?

Section 3.8: Median Hybrid Filters.

1. Show that

   (a) the 1LH+ filter preserves vertical and horizontal lines but not diagonal lines.

   (b) the R1LH+ filter preserves diagonal lines but not vertical and horizontal lines.

   (c) the 2LH+ filter preserves vertical, horizontal and diagonal lines.

2. Compare the outputs of the standard median filter and the basic median hybrid filter in the signal shown in Figure 5.3. Let $k = 5$ in the filters.

Figure 5.3. Impulse near the edge.

Section 3.9: Edge-Enhancing Selective Filters.

1. Show that the median filter preserves any monotonic degradation of a step edge.

2. Show that the comparison and selection filter has a zero-valued sequence as its impulse response for every j.

3. Show that the selective average (median) filter has a zero-valued sequence as its impulse response.

4. Show that the comparison and selection filter and the selective average (median) filter preserve step edges. What should we assume in order to guarantee that the Hachimura-Kuwahara filter will preserve step edges?

Section 3.10: Rank Selection Filters.

1. Show that permutation filters and RCRS filters are invariant to scaling by an increasing function $h(\cdot)$.

2. Show that the LUM smoother with parameter $s, 1 \leq s \leq k+1$ is identical to a center weighted median filter. Give the relation between s and the central weight.

3. Show that the LUM filter removes impulses if $s \neq 1$.

4. Show that the LUM smoother, LUM sharpener, and LUM filter all preserve step edges.

5. Show that the two definitions (3.11) and (3.12) are equivalent.

Section 3.11: M-Filters.

1. Find the ψ-functions corresponding to the mean and the median filters.

2. Show that an M-filter is scale invariant if $\psi(x)$ is linear. Can you find any example of a scale invariant M-filter where $\psi(x)$ is nonlinear?

3. Work out (3.15) from (3.14).

4. Show that the weighted mean

$$\theta = \frac{\sum_{i=1}^{N} w_i X_i}{\sum_{i=1}^{N} w_i}$$

minimizes the weighted L_2 norm

$$\sum_{i=1}^{N} w_i |X_i - \theta|^2$$

and the weighted median minimizes the weighted L_1 norm

$$\sum_{i=1}^{N} w_i |X_i - \theta|.$$

5. Assume that $\gamma > 0$ and let $\hat{\theta}$ be the the value minimizing

$$\sum_{i=1}^{N} |X_i - \theta|^{\gamma}.$$

  (a) Show that $\hat{\theta}$ is the maximum likelihood estimate of the location parameter β on a random sample $\{X_1, X_2, \ldots, X_N\}$ from a population with the exponential type density $f(x) = \alpha \exp\{-|x - \beta|^{\gamma}\}$, where α is the necessary scaling factor and $\gamma > 0$.

  (b) Show that if $\gamma > 1$ then $\hat{\theta}$ is unique.

6. Show that the modified trimmed mean filter and the modified nearest neighbor filter are actually one-step versions of M-estimators. (Hint: Use the given method of finding the output of the M-filter.)

Section 3.12: R-Filters.

1. Show that the two definitions (3.17) and (3.18) are equivalent.

2. Evaluate the number of averages from which the median is calculated if instead of pairwise averages m-wise averages would be used in the same way as in the Wilcoxon filter.

3. For what values of N and r (t) do the rank (time) Winsorized Wilcoxon filters remove impulses?

4. More weight can be given to the samples in the Wilcoxon filtering by duplicating Walsh averages $W_{i,i}$. Can we improve the step edge response by using this modification?

Section 3.13: Weighted Majority with Minimum Range Filters.

1. Why are we requiring $m > k$?

2. Study the step response of the Weighted Majority with Minimum Range Filters for the case $m > k + 1$.

Section 3.14: Nonlinear Mean Filters.

1. Assume that $a_i = 1$ for all i. Show that the geometric mean can also be written in the form of $\sqrt[N]{X_1 X_2 \cdots X_N}$.

2. Consider nonlinear mean filtering with window size 7 and assume that 6 signal samples have value a and one is an impulse $a + 100$. Does a – the value of the constant region corrupted by an impulse – have any effect on how much of the power of the impulse is seen after filtering? In other words, is $Y - a$, where Y is the output, dependent on a, and if it is, to what extent?

3. For what purpose could the weights in the definition of the nonlinear mean filter be used and how?

Section 3.15: Stack Filters.

1. Find the positive Boolean function corresponding to the 1st and Nth ranked-order filters.

2. Show that the Boolean function $f(x_{-1}, x_0, x_1) = x_{-1}x_0 + x_{-1}x_1 + x_0x_1$ corresponds to the three-point median filter.

3. Use an example to show that the stacking property does not hold for non-positive Boolean functions.

4. Show that the class of weighted order statistic filters is a subclass of the stack filter class but the class of RCRS filters is not.

5. Consider any monotonic edge

$$0, \ldots, 0, S_1 \leq S_2 \leq \cdots \leq S_k, S_k, \ldots, S_k,$$

where $0 < S_k \leq M - 1$. Show that any stack filter preserves the form of the monotonic edge but may cause it to shift.

Section 3.16: Generalizations of Stack Filters.

1. Show that the functions in (3.26) form a stacking set.

2. Show that the real domain threshold Boolean filter corresponding to the nonpositive Boolean function

   - $\bar{x}_1 x_2$ is $\text{MAX}\{0, X_2 - X_1\}$;
   - $x_1 \oplus x_2$ is $|X_1 - X_2|$.

3. One way to implement a threshold Boolean filter is to store the ordering-output table of it, in other words, a table which shows that for the ordering $X_1 \leq X_2 \leq X_3$ the output is $X_1 - X_2$, for the ordering $X_1 \leq X_3 \leq X_2$ the output is X_2, and so on. Naturally, we have altogether $N!$ orderings. Form the ordering-output table for the function $x_1\bar{x}_2 + x_1x_3$.

Section 3.17: Morphological Filters.

1. Show that flat openings and closings preserve step edges.

2. Find an example illustrating how a gray-level opening may blur a step edge.

3. A filter $F(X_1, X_2, \ldots, X_N)$ is the dual of the filter $G(X_1, X_2, \ldots, X_N)$ if $-F(-X_1, -X_2, \ldots, -X_N) = G(X_1, X_2, \ldots, X_N)$ for all $X_1, X_2, \ldots, X_N$. Show that the dilation by B is the dual of the erosion by B, the closing by B is the dual of the opening by B, and the close-opening by B is the dual of the open-closing by B.

Section 3.18: Soft Morphological Filters.

1. Show that soft closings and soft openings are not necessarily idempotent.

2. Show that soft closings and soft openings remove impulses if $1 < r \leq |B \setminus A|$.

3. Show that soft closings and soft openings preserve step edges.

Section 3.19: Polynomial Filters.

1. Consider 2nd order Volterra filters with offset term $h_0 = 0$. Show that in order to preserve homogeneous regions with an arbitrary gray-level value a, we must require the FIR-term to satisfy

$$\sum_{i=1}^{N} h_i = 1$$

and the quadratic term to satisfy

$$\sum_{i=1}^{N} \sum_{j=1}^{N} h_{i,j} = 0.$$

2. Show that if $\mathbf{H}$ is a diagonal matrix then the output of the quadratic filter is obtained by linear filtering of the squares of the input.

3. Show that if $\mathbf{H}$ is dyadic, i.e., $h_{i,j} = h_i h_j$ then the output of the quadratic filter is the square of the output of a linear filter with an impulse response h_i.

Section 3.20: Data-Dependent Filters.

1. Show that (3.30) is obtained from (3.31) if $\sigma_s^2 = \sigma_x^2 - \sigma_n^2$. Show that this equality is true if the noise is additive stationary white noise with zero mean. Show also that from this equality it follows that $0 \leq \sigma_n^2 / \sigma_x^2 \leq 1$.

2. Assume that $X^* = X_{k+1}$ and m is the sample mean. Write the definition

$$\text{LLMMSE}(X_1, X_2, \ldots, X_N) = \frac{\sigma_s^2}{\sigma_s^2 + \sigma_n^2} X^* + (1 - \frac{\sigma_s^2}{\sigma_s^2 + \sigma_n^2}) m$$

in the form of

$$b(X_1 + X_2 + \cdots + X_k + a X_{k+1} + X_{k+2} + \cdots + X_N).$$

This indicates that the LLMMSE filter is a weighted sum, with the center sample weighted differently from the others. When is it given more weight than the other samples?

3. Consider the one-dimensional case and let the window size be N. If we take k iterations of an ACLA (MCLM) filter, how many samples in fact have some effect on the output?

4. Show that the RASF filter is actually an M-estimator. (Hint: Use the method of finding the output of the M-filter discussed in Section 3.11.)

Section 3.21: Decision-Based Filters.

1. Invent some impulse detectors. Describe situations where they may fail.

2. We mentioned that one may detect impulses and edges and use the detection results in the filtering. Mention some other concepts that might be detected and whose detection provides a piece of useful information that could be utilized in filtering.

Section 3.22: Iterative, Cascaded, and Recursive Filters.

1. Do nonlinear filters commute, i.e., is the filtering result after filtering the signal first by the filter $f(\cdot)$ and then the result by the filter $g(\cdot)$ the same than after filtering the signal first by the filter $g(\cdot)$ and then the result by the filter $f(\cdot)$?

2. Consider the 5×5 image

0	0	0	0	0
0	0	1	1	0
0	0	1	1	0
0	0	1	1	0
0	0	1	1	0

Filter the image by the 3×3 recursive median filter starting

- from the upper left corner and proceeding linewise from left to right;
- from the bottom right corner and proceeding linewise from right to left.

Compare the results.

3. Can we use threshold decomposition in recursive stack filtering?

Chapter 4: STATISTICAL ANALYSIS AND OPTIMIZATION OF NONLINEAR FILTERS

1. Show that the transformation

$$\begin{cases} y_1 = x_{(1)}, \\ y_2 = x_{(2)} - x_{(1)}, \\ y_3 = x_{(3)} - x_{(2)} \end{cases}$$

has its Jacobian equal to unity and transforms the integration region $D = \{(x_{(1)}, x_{(2)}, x_{(3)}) : 0 < x_{(1)} < x_{(2)} < x_{(3)} < 1\}$ to F:

$$\begin{cases} 0 < y_1 < 1, \\ 0 < y_2 < 1 - y_1, \\ 0 < y_3 < 1 - y_1 - y_2. \end{cases}$$

2. Let $X_1, X_2, \ldots, X_N$ be independent and identically distributed random variables with a common distribution function F and density f:

- Form the distribution function of $X_{(k)}$ directly by definition. Form then the density by differentiation;
- Form the density of $X_{(k)}$ by determining the probability element of $X_{(k)}$ and then the distribution function by integrating the density.

3. Elaborate the solution of the minimization in (2.8).

4. Let $X_1, X_2, \ldots, X_N$ be random variables and assume that there is no linear relationship between them, i.e., if $P(a_1 X_1 + a_2 X_2 + \cdots + a_N X_N = 0) > 0$ then $a_1 = a_2 = \cdots = a_N = 0$. Show that the correlation matrix of $X_1, X_2, \ldots, X_N$ is positive definite. (Hint: Consider the expectation $E\{(a_1 X_1 + a_2 X_2 + \cdots + a_N X_N)^2\}$.)

5. Let $U_1, U_2, \ldots, U_N$ be independent and uniformly distributed on $[0, 1]$. Use the properties of the Beta distribution to show that

$$E\{U_{(i)}\} = \frac{i}{N+1}$$

and the joint distribution function of two order statistics (4.14) to show that

$$E\{U_{(i)} U_{(j)}\} = \frac{ij + \min\{i, j\}}{(N+1)(N+2)}, \quad i, j = 1, 2, \ldots, N.$$

6. Let $Z_1, Z_2, \ldots, Z_N$ be i.i.d. and uniformly distributed on $[-1/2, 1/2]$. Their correlation matrix is

$$\mathbf{R} = \left[\frac{4ij + 4\min\{i, j\} - 2(N+2)(i+j) + (N+1)(N+2)}{4(N+1)(N+2)} \right].$$

Show that its inverse is

$$\begin{pmatrix} \frac{3N-2}{8N} & -\frac{1}{4} & 0 & \cdots & 0 & 0 & \frac{N+2}{8N} \\ -\frac{1}{4} & \frac{1}{2} & -\frac{1}{4} & \cdots & 0 & 0 & 0 \\ \vdots & \vdots & \vdots & \ddots & \vdots & \vdots & \vdots \\ 0 & 0 & 0 & \cdots & -\frac{1}{4} & \frac{1}{2} & -\frac{1}{4} \\ \frac{N+2}{8N} & 0 & 0 & \cdots & 0 & -\frac{1}{4} & \frac{3N-2}{8N} \end{pmatrix}.$$

7. Derive the recurrence relation

$$M(\Phi, k, N, i) = M(\Phi, k, N-1, i-1) - M(\Phi, k, N, i-1), \quad 1 \le i \le N,$$

with initial values

$$M(\Phi, k, N, 0) = \int_{-\infty}^{\infty} x^k \frac{d}{dx}(\Phi(x)^N)\, dx, \quad i = 0, 1, \ldots, N-1, \quad 0 \le N$$

from the definition of the numbers $M(\Phi, k, N, i)$ by expanding in the integral (4.39).

8. (Fairly difficult.) Derive the relation (4.46) and its inverse.

9. Show that the number of different stack filters with window size N is greater than $2^{2^N/N}$. (Hint: Consider the Boolean functions that are formed by sums of subsets of all monomials with $(N+1)/2$ variables.)

10. In Example 4.12 derive the inequality constraints from the requirements stated in the example. Show that if we also require that the filter is self-dual, then a solution is given by the weighted median filter with the weight vector $(1, 3, 3, 3, 1)$.

11. Compute the output distribution function of the multistage median of Example 4.14.

12. Show that the cascade of two median filters of length three is a weighted median filter of length five.

13. Consider the basic median hybrid filter defined in Section 3.8. Let F, H and G be the distribution functions of the outputs of F_1, F_2 and the center sample, respectively, and f, g and h be the corresponding density functions. Use formula (4.27) to show that for independent inputs the output density function of the whole filter is

$$m = f(G + H) + g(H + F) + h(F + G) - 2(fGH + gHF + hFG).$$

BIBLIOGRAPHY

[1] S. Agaian, J. Astola, and K. Egiazarian, *Binary Polynomial Transforms and Digital Filters*, Marcel Dekker, New York, 1995.

[2] D. F. Andrews, P. J. Bickel, F. R. Hampel, P. J. Huber, W. H. Rogers, and J. W. Tukey, *Robust Estimates of Location: Survey and Advances*, Princeton University Press, Princeton, N.J., 1972.

[3] Anonymous, "Dissertation sur la recherche de milieu le plus probable, entre les résultats de plusieurs observations ou expériences," *Ann. Math. Pures Appl.*, vol. 12, pp. 181-204, 1821.

[4] G. R. Arce and M. P. McLoughlin, "Theoretical analysis of the max/median filter," *IEEE Trans. Acoust., Speech, Signal Processing*, vol. ASSP-35, pp. 60–69, Jan. 1987.

[5] J. Astola, P. Heinonen, and Y. Neuvo, "On root structures of median and median-type filters," *IEEE Trans. Acoust., Speech, Signal Processing*, vol. ASSP-35, pp. 1199–1201, Aug. 1987.

[6] J. Astola, O. Yli-Harja, P. Heinonen, and Y. Neuvo, "Class of edge preserving median type filters using logical operations on linear substructures," in *Proc. Eur. Conf. Circ. Theor. Design ECCTD-87*, Paris, France, Sept. 1987, pp. 609–614.

[7] J. Astola and G. Campbell, "On computation of the running median," *IEEE Trans. Acoust., Speech, and Signal Processing*, vol. 37, pp. 572–574, April 1989.

[8] J. Astola and Y. Neuvo, "Optimal median type filters for exponential noise distributions," *Signal Processing*, vol. 17, pp. 95–104, June 1989.

[9] J. Astola, L. Koskinen, O. Yli-Harja, and Y. Neuvo, "Digital filters based on threshold decomposition and Boolean functions," in *Proc. SPIE Symp., Visual Comm., Image Proc.*, Philadelphia, USA, Nov. 1989, pp. 461–470.

[10] J. Astola, P. Heinonen, and Y. Neuvo, "Linear median hybrid filters," *IEEE Trans. Circuits and Syst.*, vol CAS-36, pp. 1430–1438, Nov. 1989.

[11] K. E. Barner and G. R. Arce, "Permutation filters: a class of nonlinear filters based on set permutations," *IEEE Trans. on Signal Processing*, vol. 42, pp. 782–798, April 1994.

[12] V. Barnett and T. Lewis, *Outliers in Statistical Data*, Wiley, Chichester, 1978.

[13] A. E. Beaton and J. W. Tukey, "The fitting of power series, meaning polynomials, illustrated in band-spectroscopic data," *Technometrics*, vol. 16, pp. 147–186, 1974.

[14] J. B. Bednar and T. L. Watt, "Alpha-trimmed means and their relationship to median filters," *IEEE Trans. Acoust., Speech, Signal Processing*, vol. ASSP-32, pp. 145–153, Feb. 1984.

[15] P. J. Bickel, "On some robust estimates of location," *Ann. Math. Statist.*, vol. 36, pp. 847-858, 1965.

[16] P. Bloomfield and W. L. Steiger, *Least Absolute Deviations: Theory, Applications, and Algorithms*, Birkhäuser, Boston, 1983.

[17] A. C. Bovik, T. S. Huang, and D. C. Munson, Jr., "Nonlinear filtering using linear combinations of order statistics," in *Proc. 1982 IEEE Int. Conf. on Acoustics, Speech and Signal Processing*, Paris, May 1982, pp. 2067–2070.

[18] A. C. Bovik, T. S. Huang, and D. C. Munson, Jr., "A generalization of median filtering using linear combinations of order statistics," *IEEE Trans. Acoust., Speech, Signal Processing*, vol. ASSP-31, pp. 1342–1350, Dec. 1983.

[19] D. R. K. Brownrigg, "The weighted median filter," *Commun. ACM*, vol. 27, pp. 807–818, Aug. 1984.

[20] D. R. K. Brownrigg, "Generation of representative members of an RrSst weighted median filter class," *IEE Proc. Part. F, Commun., Radar, Signal Process.*, vol. 133, pp. 445–448, Aug. 1986.

[21] T. H. Cormen, C. E. Leiserson, and R. L. Rivest, *Algorithms*, The MIT Press, Cambridge, Massachusetts, 1994.

[22] A. T. Craig, "On the distribution of certain statistics," *Amer. J. Math.*, vol. 54, pp. 353–366, 1932.

[23] R. J. Crinon, "The Wilcoxon filter: a robust filtering scheme," in *Proc. IEEE Int. Conf. Acoust., Speech, and Signal Processing 85*, Tampa, March 1985, pp. 668–671.

[24] L. S. Davis and A. Rosenfeld, "Noise cleaning by iterated local averaging," *IEEE Trans. Syst. Man Cybern.*, vol. SMC-8, pp. 705-710, Sept. 1978.

[25] P. P. Gandhi and S. A. Kassam, "Performance of some rank filters for edge preserving smoothing," in *Proc. 1987 IEEE Symposium on Circuits and Systems*, Philadelphia, May 1987, pp. 264-267.

[26] P. P. Gandhi, I. Song, and S. A. Kassam, "Nonlinear smoothing filters based on rank estimates of location," *IEEE Trans. Acoust., Speech, Signal Processing*, vol. 37, pp. 1359-1379, Sept. 1989.

[27] C. F. Gauss, "Bestimmung der genauigkeit der Beobachtungen," *Z. Astr. verw. Wiss*, 1816. In *Werke 4*, Dieterichsche Universitäts-Druckerei, Göttingen, pp. 109–117, 1880.

[28] E. P. Gilbo and I. B. Chelpanov, *Obrabotka Signalov na Osnove Uporyadochennogo Vybora* (Signal Processing Based on Ordering), Izd. Sovetskoe Radio, Moscow, 1976. (In Russian.)

[29] F. R. Hampel, "Beyond location parameters: robust concepts and methods (with discussion)," in *Proc. of the 40th Session of the ISI*, XLVI, Book 1, 1975, pp. 375-391.

[30] F. R. Hampel, E. M. Ronchetti, P. J. Rousseeuw, and W. A. Stahel, *Robust Statistics: The Approach Based on Influence Functions*, Wiley, New York, 1986.

[31] R. C. Hardie and C. G., Jr. Boncelet, "LUM filters: a class of rank order based filters for smoothing and sharpening," *IEEE Trans. Signal Processing*, vol. 41, pp. 1061–1076, March 1993.

[32] R. C. Hardie and K. E. Barner, "Rank conditioned rank selection filters for signal restoration," *IEEE Trans. on Image Processing*, vol. 3, pp. 192-206, March 1994.

[33] R. C. Hardie and K. E. Barner, "Extended permutation filters and their application to edge enhancement," *IEEE Trans. on Image Processing*, vol. 5, pp. 855-867, June 1996.

[34] H. L. Harter, "The method of least squares and some alternatives," Parts I-VI, *Rev. Int. Inst. Statist.*, vol. 42, pp. 147–174, 1974 (Part I), vol. 42, pp. 235–264, 1974 (Part II), vol. 43, pp. 1–44, 1975 (Part III), vol. 43, pp. 125–190, 1975 (Part IV), vol. 43, pp. 269–278, 1975 (Part V), vol. 44, pp. 113–159, 1976 (Part VI).

[35] H. J. A. M. Heijmans, *Morphological Image Operators*, Academic Press, Boston, 1994.

[36] P. Heinonen and Y. Neuvo, "Smoothed median filters with FIR substructures," in *Proc. 1985 IEEE Int. Conf. Acoust., Speech, and Signal Processing*, Tampa, March 1985, pp. 49–52.

[37] J. L. Hodges, Jr. and E. L. Lehmann, "Estimates of location based on rank tests," *Ann. Math. Statist.*, vol. 34, pp. 598-611, 1963.

[38] T. S. Huang, G. J. Yang, and G. Y. Tang, "A fast two-dimensional median filtering algorithm," *IEEE Trans. Acoust., Speech, Signal Processing*, vol. ASSP-27, pp. 13–18, Febr. 1979.

[39] P. J. Huber, "Robust estimation of a location parameter," *Ann. Math. Statist.*, vol. 35, pp. 73–101, 1964.

[40] P. J. Huber, *Robust Statistics*, Wiley, New York, 1981.

[41] H. Huttunen, P. Kuosmanen, and J. Astola, "Optimal RASF Filtering," in *Proc. of Nonlinear Image Processing VII, SPIE Proceedings*, vol. 2662, San Jose, Febr. 1996, pp. 154-165.

[42] L. A. Jaeckel, *Robust Estimates of Location*, Ph. D. Thesis, Princeton University, 1951.

[43] A. K. Jain, *Fundamentals of Digital Image Processing*, Prentice-Hall, Englewood Cliffs, New Jersey, 1989.

[44] B. I. Justusson, "Median filtering: Statistical properties," *Topics in Applied Physics, Two-Dimensional Digital Signal Processing II*, Huang, T. S., Ed., Springer-Verlag, Berlin, vol. 43, pp. 161–196, 1981.

[45] S. A. Kassam and S. R. Peterson, "Nonlinear finite moving window filters for signal restoration," in *Proc. IEEE Pacific RIM Conf. on Comm. Computer, and Signal Processing*, Victoria, Canada, June 1987, pp. 17-20.

[46] S. Kay, *Fundamentals of Statistical Signal Processing*, Prentice-Hall, Englewood Cliffs, New Jersey, 1993.

[47] B. G. Kendall, *The Advanced Theory of Statistics*, vol. 1, Charles Griffin & Company Limited, London, 1973.

[48] D. E. Knuth, *The Art of Computer Programming*, vol. 3, Addison-Wesley, Reading, MA, 1973.

[49] S.-J. Ko, Y. H. Lee, and A. T. Fam, "Selective median filters," in *Proc. 1988 IEEE Symposium on Circuits and Systems*, Helsinki, June 1988, pp. 1495-1498.

[50] S.-J. Ko and Y. H. Lee, "Center weighted median filters and their applications to image enhancement," *IEEE Trans. Circuits and Systems*, vol. 38, pp. 984–993, Sept. 1991.

[51] L. Koskinen, J. Astola and Y. Neuvo, "Soft morphological filters," in *Proc. SPIE Symp. on Image Algebra and Morphological Image Processing II*, San Diego, July 1992, pp. 262–270.

[52] D. T. Kuan, A. A. Sawchuk, T. C. Strand, and P. Chavel, "Adaptive noise smoothing filter for images with signal-dependent noise," *IEEE Trans. Pattern Anal. Mach. Intell.*, vol. PAMI-7, pp. 165-177, March 1985.

[53] A. Kundu, S. K. Mitra, P. P. and Vaidyanathan, "Generalized mean filters: a new class of nonlinear filters for image processing," in *Proc. of 6th Symp. Circuit Theory Design*, Stuttgart, Sept 1983, pp. 185-187.

[54] A. Kundu and W.-R. Wu, "Double-window Hodges-Lehmann (D) filter and hybrid D-median filter for robust image smoothing," *IEEE Trans. Acoust., Speech, Signal Processing*, vol. 37, pp. 1293-1298, Aug. 1989.

[55] P. Kuosmanen, L. Koskinen, and J. Astola, "An adaptive morphological filtering method," in *Proc. EUSIPCO -92*, Brussels, Aug. 1992, pp. 1089-1092.

[56] P. Kuosmanen, *Statistical analysis and optimization of stack filters*, Acta Polytechnica Scandinavica, Electrical Engineering Series, 77, Helsinki, 1994.

[57] M. Kuwahara, K. Hachimura, S. Eiho, and M. Kinoshita, "Processing of RI-angiocardiographic image," in *Digital Processing of Biomedical Images*, Preston, K. and Onoe, M., Eds., Plenum, New York, 1976, pp. 187-202.

[58] C. Lantuéjoul and J. Serra, "*M*-Filters," in *Proc. 1982 IEEE Int. Conf. on Acoustics, Speech and Signal Processing*, Paris, May 1982, pp. 2063-2066.

[59] J.-H. Lee and J.-Y. Kao, "A fast algorithm for two-dimensional Wilcoxon filtering," in *Proc. 1987 IEEE Symposium on Circuits and Systems*, Philadelphia, May 1987, pp. 268-271.

[60] J.-S Lee, "Digital image enhancement and noise filtering by use of local statistic," *IEEE Trans. Pattern Anal. Mach. Intell.*, vol. PAMI-2, pp. 165-168, March 1980.

[61] J.-S. Lee, "Digital image smoothing and the sigma filter," *Computer Vision, Graphics, and Image Processing*, vol. 24, pp. 255-269, Nov. 1983.

[62] K. D. Lee and Y. H Lee, "Threshold Boolean filters," *IEEE Trans. on Signal Processing*, vol. 42, pp. 2022-2036, Aug. 1994.

[63] Y. H. Lee and S. A. Kassam, "Some generalizations of median filters," in *Proc. 1983 IEEE Int. Conf. Acoust., Speech, and Signal Processing*, Boston, April 1983, pp. 411-414.

[64] Y. H. Lee and S. A. Kassam, "Generalized median filtering and related nonlinear filtering techniques," *IEEE Trans. Acoust., Speech, Signal Processing*, vol. ASSP-33, pp. 672-683, June 1985.

[65] Y. H. Lee and A. T. Fam, "An edge gradient enhancing adaptive order statistic filter," *IEEE Trans. Acoust., Speech, Signal Processing*, vol. ASSP-35, pp. 680-695, May 1987.

[66] E. L. Lehmann, *Theory of Point Estimation*, Wiley, New York, 1983.

[67] A. Lev, S. W. Zucker, and A. Rosenfeld, "Iterative enhancement of noisy images," *IEEE Trans. Syst. Man Cybern.*, vol. SMC-7, pp. 435-442, June 1977.

[68] J. H. Lin and E. J. Coyle, "Minimum mean absolute error estimation over the class of generalized stack filters," *IEEE Trans. on Acoust., Speech, and Signal Processing*, vol. 38, pp. 663-678, April 1990.

[69] L.-B. Ling, R. Yin, and X.-H. Wang, "Nonlinear filters for reducing spiky noise: 2-dimension," in *Proc. 1984 IEEE Int. Conf. Acoust., Speech, and Signal Processing*, San Diego, March, 1984, pp. 31.8.1–31.8.4.

[70] E. H. Lloyd, "Least-square estimation on location and scale parameters using order statistics," *Biometrika*, vol. 39, pp. 88–95, 1952.

[71] H. G. Longbotham, A. C. Bovic, and A. Restrepo, "Generalized order statistic filters," in *Proc. 1989 IEEE Int. Conf. Acoust., Speech, and Signal Processing*, Glasgow, May 1989, pp. 1610–1613.

[72] H. G. Longbotham and D. Eberly, "The WMMR filters: a class of robust edge enhancers," *IEEE Trans. on Signal Processing*, vol. 41, pp. 1680-1685, April 1993.

[73] C. L. Mallows, "Some theory of nonlinear smoothers," *The Annals of Statistics*, vol. 8, pp. 695–715, 1980.

[74] C. L. Mallows, "Resistant smoothing," *Time Series*, Anderson, O. D., Ed., North-Holland, Amsterdam, pp. 147-155, 1980.

[75] P. Matheron, *Random Sets and Integral Geometry*, Wiley, New York, 1975.

[76] D. I. Mendeleev, "Course of work on the renewal of prototypes or standard measures of lengths and weights", *Vremennik Glavnoi Palaty Mer i Vesov*, vol. 2, pp. 157–185, 1895, (in Russian). Reprinted: *Collected writings (Socheneniya)*, Izdat. Akad. Nauk, SSSR, Leningrad-Moscow, vol. 22, pp. 175–213, 1950.

[77] S. Muroga, *Threshold Logic and Its Applications*, Wiley, New York, 1971.

[78] P. M. Narenda, "A separable median filter for image noise smoothing," *IEEE Trans. Pattern Anal. Mach. Intell.*, vol. PAMI-3, pp. 20-29, Jan. 1981.

[79] A. Nieminen, P. Heinonen, and Y. Neuvo, "A new class of detail-preserving filters for image processing," *IEEE Trans. Pattern Anal. Mach. Intell.*, vol. PAMI-9, pp. 74–90, Jan. 1987.

[80] A. Nieminen and Y. Neuvo, "Comments of theoretical analysis of the max/median filter," *IEEE Trans. on Acoust., Speech, and Signal Process.*, vol. ASSP-36, pp. 826–827, May 1988.

[81] J. Niewęgłowski, M. Gabbouj, and Y. Neuvo, "Weighted medians—positive Boolean function conversion," *Signal Processing*, vol. 34, pp. 149–162, Nov. 1993.

[82] T. A. Nodes and N. C. Gallagher, Jr., "Median filters: some modifications and their properties," *IEEE Trans. on Acoust., Speech, and Signal Processing*, vol. ASSP-30, pp. 739-746, Oct. 1982.

[83] A. V. Oppenheim, R. W. Schafer, and T. G. Stockham, Jr., "Nonlinear filtering of multiplied and convolved signals," *Proc. of IEEE*, vol. 56, pp. 1264–1291, Aug. 1968.

[84] F. Palmieri and C. G. Boncelet, "*Ll*-filters—a new class of order statistic filters," *IEEE Trans. Acoust., Speech, Signal Processing*, vol. 37, pp. 691–701, May 1989.

[85] N. Pappas and I. Pitas, "Greyscale morphology using nonlinear L_p mean filters," in *Proc. 1995 IEEE Workshop on Nonlinear Signal and Image Processing*, vol. I, Neos Marmaras, June 1995, pp. 34–37.

[86] S. R. Peterson and S. A. Kassam, "Edge preserving signal Enhancement using generalizations of order statistic filters," in *Proc. IEEE Int. Conf. Acoust., Speech, and Signal Processing 85*, Tampa, March 1985, pp. 672–675.

[87] B. Picinbono, "Quadratic filters," in *Proc. 1982 IEEE Int. Conf. on Acoustics, Speech and Signal Processing*, Paris, May 1982, pp. 298–301.

[88] I. Pitas and A. N. Venetsanopoulos, "Nonlinear mean filters in image processing," *IEEE Trans. on Acoust., Speech, and Signal Processing*, vol. ASSP-34, pp. 573–584, June 1986.

[89] I. Pitas and A. N. Venetsanopoulos, *Nonlinear Digital Filters: Principles and Applications*, Kluwer Academic Publishers, Boston, 1990.

[90] C. C. Pu and F. Y. Shih, "Soft mathematical morphology: binary and gray scale," in *Proc. International Workshop on Mathematical Morphology and Its Applications to Signal Processing*, Barcelona, May 1993, pp. 28–33.

[91] G. Ramponi and G. L. Sicuranza, "Quadratic filters for image processing," *IEEE Trans. on Acoust., Speech, and Signal Processing*, vol. 36, pp. 937-939, June 1988.

[92] G. Ramponi, G. L. Sicuranza, and W. Ukovich, "A computational method for the design of 2-D nonlinear Volterra filters," *IEEE Trans. on Circuits and Systems*, vol. 35, pp. 1095-1102, Sept. 1988.

[93] P. J. Rousseeuw, "Least median of squares regression," *J. Amer. Statist. Assoc.*, vol. 79, pp. 871-880, 1984.

[94] D. E. Rutherford, *Introduction to Lattice Theory*, Oliver and Boyd, Essex, U.K., 1965.

[95] J. Serra, *Image Analysis and Mathematical Morphology*, Academic Press, London, 1982.

[96] J. Serra, ed., *Image Analysis and Mathematical Morphology, Part II: Theoretical Advances*, Academic Press, London, 1988.

[97] G. L., Sicuranza, "Quadratic filters for signal processing," *Proc. of the IEEE*, vol. 80, pp. 1263–1285, Aug. 1992.

[98] S. R. Sternberg, "Grayscale morphology," *Computer Vision, Graphics, and Image Processing*, vol. 35, pp. 333-355, 1986.

[99] G. Strang, *Linear Algebra and its Applications*, Harcourt Brace Jovanovich College Publishers, Orlando, 1988.

[100] C. Sun and P. D. Rabinowitz, "Recursive approaching signal filter," *IEEE Signal Processing Letters*, vol. 2, pp. 85–88, May 1995.

[101] X. Z. Sun and A. N. Venetsanopoulos, "Adaptive schemes for noise filtering and edge detection by use of local statistics," *IEEE Trans. Circuits and Systems*, vol. 35, pp. 57–69, Jan. 1988.

[102] I. Tabus, M. Gabbouj, and L. Yin, "Real domain adaptive WOS filtering using neural network approximations," in *Proc. IEEE Winter Workshop on Nonlinear Digital Signal Processing*, Tampere, Jan. 1993, pp. 7.2-1.1–7.2-1.6.

[103] J. W. Tukey, "A survey of sampling from contaminated distributions," in *Contributions to Probability and Statistics*, Olkin, I., Ed., Stanford Univ. Press, pp. 448–485, Stanford, 1960.

[104] J. W. Tukey, "Nonlinear (nonsuperposable) methods for smoothing data," in *Congr. Rec. EASCON -74*, p. 673, 1974. (Abstract only.)

[105] J. W. Tukey, *Exploratory Data Analysis*, Addison-Wesley, Reading, MA, 1977 (1970-71: preliminary edition).

[106] J. W. Tukey, "The ninther, a robust estimate of location," in *Contributions to Survey Sampling and Applied Statistics*, David, H. A., Ed., Academic, New York, 1978, ???PAGES???.

[107] S. G. Tyan, "Median filtering: Deterministic properties," *Topics in Applied Physics, Two-Dimensional Digital Signal Processing II*, Huang, T. S., Ed., Springer-Verlag, Berlin, vol. 43, pp. 197–217, 1981,

[108] V. Volterra, Sopra le funzioni che dipendono da altre funzioni, Rend. Regia Accademia dei Lincei, 2o Sem., pp.97-105, 141-146, 153-158, 1887.

[109] V. Volterra, *Leçons Sur Les Fonctions De Lignes*, Gauthier-Villars, Paris, 1913.

[110] V. Volterra, *Theory of Functionals and of Integral and Integro-Differential Equations*, Dover, New York, 1959.

[111] P. D. Wendt, E. J. Coyle, and N. C. Gallagher, Jr., "Stack filters: their definition and some initial properties," in *Proc. 1985 Conf. Inform. Sci. Syst.*, Baltimore, MD, March 1985, pp. 378–383.

[112] P. D. Wendt, E. J. Coyle, and N. C. Gallagher, Jr., "Stack filters," *IEEE Trans. on Acoust., Speech, and Signal Processing*, vol. ASSP-34, pp. 898-911, Aug. 1986.

[113] N. Wiener, *Nonlinear Problems in Random Theory*, The Technology Press, MIT and John Wiley and Sons, New York, 1958.

[114] R. Wichman, J. Astola, P. Heinonen, and Y. Neuvo, "FIR-median hybrid filters with excellent transient response in noisy conditions," *IEEE Trans. Acoust., Speech, Signal Processing*, vol. 38, pp. 2108–2117, Dec. 1990.

[115] R. Wichman and Y. Neuvo, "Multilevel median filters for image processing," in *Proc. 1991 IEEE Symposium on Circuits and Systems*, Singapore, June 1991, pp. 412–415.

[116] S. S. Wilks, *Mathematical Statistics*, John Wiley & Sons Inc., New York, 1962.

[117] R. Yang, L. Yin, M. Gabbouj, J. Astola, and Y. Neuvo, "Optimal weighted median filters under structural constraints," *IEEE Trans. Signal Processing*, vol. 43, pp. 591–604, March 1995.

[118] L. Yin and Y. Neuvo, "Adaptive FIR-WOS hybrid filtering," in *Proc. 1992 IEEE Symposium on Circuits and Systems*, San Diego, May 1992, pp. 2637–2640.

[119] L. Yin and Y. Neuvo, "Fast adaptation and performance characteristics of FIR-WOS hybrid filters," *IEEE Trans. Signal Processing*, vol. 42, pp. 1610–1628, May 1993.

[120] L. Yin, R. Yang, M. Gabbouj, and Y. Neuvo, "Weighted median filters: a tutorial," *IEEE Trans. on Circuits and Systems—II: Analog and Digital Signal Processing*, vol. 43, pp. 157–192, March 1996.

[121] O. Yli-Harja, J. Astola, and Y. Neuvo, "Analysis of the properties of median and weighted median filters using threshold logic and stack filter representation," *IEEE Trans. Signal Processing*, vol. 39, pp. 395-410, Febr. 1991.

[122] O. Yli-Harja, "Formula for the joint distribution of stack filters," *Signal Processing Letters*, vol. 1, pp. 129-130, Sept. 1994.

Index

9 780367 448257